Ali Caglar Özen

Novel MRI Technologies for Structural and Functional Imaging of Tissues with Ultra-short T_2 Values

Schriften des Instituts für Mikrostrukturtechnik
am Karlsruher Institut für Technologie (KIT)
Band 34

Hrsg. Institut für Mikrostrukturtechnik

Eine Übersicht aller bisher in dieser Schriftenreihe
erschienenen Bände finden Sie am Ende des Buchs.

Novel MRI Technologies for Structural and Functional Imaging of Tissues with Ultra-short T_2 Values

von
Ali Caglar Özen

KIT Scientific Publishing

Dissertation, Karlsruher Institut für Technologie
KIT-Fakultät für Maschinenbau

Tag der mündlichen Prüfung: 30. Januar 2017
Hauptreferent: Prof. Dr. Jan G. Korvink
Korreferent: Prof. Dr. Michael Bock

Impressum

KIT Scientific Publishing

Karlsruher Institut für Technologie (KIT)
KIT Scientific Publishing
Straße am Forum 2
D-76131 Karlsruhe

www.ksp.kit.edu

Print on Demand 2017 – Gedruckt auf FSC-zertifiziertem Papier

ISSN 1869-5183
ISBN 978-3-7315-0657-7
DOI 10.5445/KSP/1000068880

Novel MRI Technologies for Structural and Functional Imaging of Tissues with Ultra-short T_2 Values

Zur Erlangung des akademischen Grades

Doktor der Ingenieurwissenschaften

der Fakultät für Maschinenbau
Karlsruher Institut für Technologie (KIT)
genehmigte

Dissertation

von

M.Sc. Ali Caglar Özen

Tag der mündlichen Prüfung: 30. Januar 2017
Hauptreferent: Prof. Dr. Jan G. Korvink
Korreferent: Prof. Dr. Michael Bock

Sevalim'e…

Abstract

Typically, a Nuclear Magnetic Resonance (NMR) or Magnetic Resonance Imaging (MRI) experiment is composed of signal excitation, spectral or spatial encoding, and detection of the resulting signal. The encoding block is used to resolve the spatial distribution or other functional parameters such as motion. In general, there are two strategies to generate signal: concurrent and interleaved excitation and acquisition. Concurrent excitation and acquisition (CEA) was broadly employed in the first two decades of NMR. In fact, the first MR image acquisition by Lauterbur was based on CEA. Triggered by the technological advances, in the 1970s CEA was replaced by interleaved acquisition methods in MRI because of better signal quality and efficiency. Since then, MRI research is focused on improving the interleaved acquisition techniques, and by today, all standard hardware and software tools use interleaved methods.

Conventional MRI has several fundamental limitations such as long scan durations, artifacts in images caused by subject motion, very high acoustic noise levels, signal loss due to short relaxation times, and RF induced heating of electrically conducting objects. The goals of this thesis are to evaluate state-of-the-art methods for imaging of tissue with short relaxation times, to prove the feasibility of CEA in a clinical MRI system, and to introduce a new electrophysiological measurement unit applied simultaneously with dynamic lung MRI. All these developments are targeted towards improving MRI of short-T_2* tissues such as lung tissue or mummified samples. In particular, CEA offers 100% signal acquisition efficiency, shorter scan time, zero acquisition delay, low peak RF power, and silent scan. Furthermore, CEA features a zero acquisition delay that allows capturing the signal of fast-decaying spins, while this signal is largely lost in conventional methods. Finally, CEA pulse sequences do not demand high-peak power RF pulses and rapidly switching strong gradient fields.

This thesis starts with assessing the limits of state-of-the-art pulse sequences for imaging of ancient mummified samples. MRI of mummies is challenging due to the extremely low water content and the very short transverse relaxation time, T_2. The objective of this study is to compare 3 different short-TE pulse sequences: ultra-short echo time (UTE), point-wise encoding time reduction with radial acquisition (PETRA), and single point imaging (SPI). MR images of an ancient mummified human hand were acquired with all 3 sequences at field strengths of 1.5 T and 3 T using home-made transmit/receive (Tx/Rx) RF resonators. Based on the obtained contrast, different types of tissues were identified in all images, and anatomical structures such as bones and tendons were clearly separated. Quantitative relaxation time mapping was possible for tissue components with T_2* higher than 100 μs. This work provided an initial analysis for the optimization of imaging protocols to define an MRI-standard of reference for paleoradiology studies, and ultimately for MRI of tissue with extremely short T_2.

In the second part, CEA was implemented on a clinical MRI system which has been considered impossible so far. Therefore, a novel decoupling method was developed to isolate the transmit (Tx) and receive (Rx) resonators. The main Tx and Rx resonators were geometrically decoupled to reduce mutual inductance (isolation of 20 dB). The remaining signal that was induced by the main Tx into the Rx was reduced further by application of a second field, whose phase and amplitude were adjusted to interfere destructively with the unwanted signal of the main Tx (isolation of 50 dB). Isolation performance and transmit noise behavior were analyzed for different loading conditions, and a CEA MRI experiment was performed in a phantom with sub-millisecond T_2* and in an animal *ex vivo*. This study demonstrated the feasibility of CEA imaging experiment in clinical MRI systems with minor hardware modifications using active decoupling with parallel transmit array capabilities.

Another novel decoupling method for CEA was realized using a fully automated analog cancellation unit. The cancellation circuit was composed of a voltage-controlled phase shifter and attenuator pair, which changes phase and amplitude of a tapped small copy of the Tx signal in a real-time feedback from MRI system. A gradient descent algorithm was implemented to find the phase and attenuation settings that suppress the Tx leakage most. Residual Tx leakage was subtracted during reconstruction based on the reference measurements of a pick-up loop detector. The automated analog cancellation system was tested in a 3T clinical MRI system, and CEA MRI of a phantom and an *in vivo* human wrist were done to demonstrate the feasibility of the concept.

Finally, another novel system was developed for simultaneous vocal tract measurements and dynamic MRI of the lungs to provide fundamental insights into the lung physiology during singing. To analyze vocal fold oscillatory patterns during dynamic lung MRI, an electroglottography (EGG) system was modified to allow for simultaneous EGG measurements during MR image acquisitions. A low-pass filter was introduced to suppress residual RF coupling into the EGG signal. RF heating was tested in a gel phantom to ensure MR safety, and functionality of the device was assessed in a volunteer experiment at singing frequencies from A5 to A3. In the recorded EGG signals, remaining RF interferences were removed by ICA post processing, and standard EGG parameters such as fundamental frequency, contact quotient and jitter were calculated. In a second volunteer experiment, EGG recordings were compared with lung diameter measurements from 2D time-resolved acquisitions. This study shows that EGG recording is possible during dynamic lung MRI of professional singers, and glottal activity can be studied safely at 1.5 T.

Zusammenfassung

Ein Kernspin- (NMR) oder Magnetresonanztomographie-(MRT)-Experiment besteht typischerweise aus einer Signalanregung, einer spektralen oder räumlichen Kodierung und der Detektion des resultierenden Signals. Der Kodierteil wird dazu verwendet, die räumliche Verteilung oder andere funktionelle Parameter wie die Geschwindigkeit zu messen. Im Allgemeinen gibt es zwei Strategien für die Signalerzeugung: eine simultane oder eine abwechselnde Anregung und Auslese. Die simultane Anregung und Auslese (concurrent excitation and acquisition, CEA) wurde vielfach verwendet in den ersten beiden Jahrzehnten der NMR; so beruht beispielsweise die erste MR Bildaufnahme durch Lauterbur auf einer CEA Technik. Angestoßen durch technische Verbesserungen wurde CEA in den 1970er Jahren in der MRT durch abwechselnde Akquisitionsmethoden ersetzt, da hiermit eine bessere Signalqualität und eine höhere Effizienz erzielt werden konnten. Seit dieser Zeit ist die MRT-Forschung darauf fokussiert, die abwechselnden Akquisitionstechniken zu verbessern, so dass heute alle Standardsysteme abwechselnde Techniken einsetzen.

Die konventionelle MRT hat einige fundamentale Limitationen wie lange Aquisitionszeiten, Bildartefakte durch Bewegung, hohe Geräuschpegel, Signalverlust bei kurzen Relaxationszeiten und HF-induzierte Erhitzungen in elektrisch leitfähigen Objekten. Die Ziele dieser Arbeit sind es, etablierte Bildgebungsmethoden zur Darstellung von Geweben mit kurzen Relaxationszeiten zu evaluieren, die generelle Durchführbarkeit von CEA-Messungen an klinischen MRT-Systemen zu zeigen, und ein neues elektrophysiologisches Meßsystem vorzustellen, das während dynamischer Lungen-MRT-Aufnahmen verwendet werden kann. All diese Entwicklungen zielen darauf ab, die MRT von Geweben zu verbessern, die wie Lungengewebe oder mumifizierte Proben ein kurzes T_2* besitzen. Im Speziellen bietet hierzu CEA eine 100% Effizienz der Signalakquisition, kürzere Messzeiten, verschwindende

Totzeiten, geringe Spitzensendeleistung, sowie geräuscharme Untersuchungen. Zusätzlich weist CEA eine verschwindende Totzeit auf, so dass auch schnell zerfallende Signale erfasst werden können, die mit konventionellen Methoden nicht detektierbar sind. Schließlich benötigen CEA Pulssequenzen auch keine HF-Pulse mit hohen Spitzenleistungen oder schnell geschaltete Gradientenfelder.

In dieser Arbeit werden zuerst die Grenzen von bestehenden Pulssequenzen zur Bildgebung von historischen, mumifizierten Proben ausgeleuchtet. Die MRT von Mumien ist schwierig wegen ihres extrem geringen Wassergehalts und der sehr kurzen transversalen Relaxationszeiten T_2. Das Ziel dieser Studie ist es, 3 verschiedene Pulssequenzen mit kurzem TE zu vergleichen: ultra-short echo time (UTE), point-wise encoding time reduction with radial acquisition (PETRA), und single point imaging (SPI). MRT-Bilder eine historischen mumifizierten Hand wurden mit allen drei Sequenzen bei Feldstärken von 1.5 T und 3 T unter Einsatz von selbstentwickelten Sende-Empfangs-(Tx/Rx)-spulen aufgenommen. Auf der Grundlage der erzielten Kontraste wurden in allen Bildern verschiedene Gewebearten identifiziert, und anatomische Strukturen wie Knochen oder Bänder voneinander unterschieden. Eine quantitative Kartierung der Relaxationszeit konnte für alle Gewebekomponenten mit einem T_2* von mehr als 100 μs erreicht werden. Diese Arbeit ist damit eine erste Analyse zur Optimierung von Bildgebungsprotokollen als Referenzstandard für die Paleoradiologie, mit dem Ziel, die MRT auch bei Geweben mit extrem kurzem T_2 einsetzen zu können.

Im zweiten Teil der Arbeit wurde CEA an einem klinischen MRT-System implementiert – dies wurde bisher als unmöglich erachtet. Eine neue Entkopplungsmethode wurde entwickelt, um die Sende- (Tx) von den Empfangs-(Rx)-resonatoren zu isolieren. Die Tx und Rx Resonatoren wurden geometrisch entkoppelt um die Gegeninduktivität zu reduzieren (Isolation: 20 dB). Das verbleibende Signal, das durch die Sendespule im Empfangskanal induziert wurde, konnte durch den Einsatz eines zweiten Sendefeldes weiter reduziert werden, dessen Phase und Amplitude so justiert wurden, dass es zu einer destruktiven Interferenz mit dem Signal der Hauptsendespule kam (Isolation: 50 dB). Die Isolationsstärke und des Rauschverhalten des Senders wurden analysiert unter verschiedenen Beladungssituationen, und ein CEA MRT Experiment wurde durchgeführt sowohl in einem Phantom mit Sub-Millisekunden T_2* als auch in einem Tierkadaver. Mit dieser Studie konnte erstmals die Durchführbarkeit von CEA Bildgebungsexperimenten auf klinischen MRT-Systemen gezeigt werden, wobei die aktive Entkopplung mit einer parallelen Sendeeinheit realisiert wurde.

Eine weitere Entkopplungsmethode für CEA wurde realisiert mit Hilfe eines vollständig automatisierten analogen Unterdrückungsschaltkreises (cancellation unit). Der Schaltkreis bestand aus spannungsgesteuerten Phasenschieber und Dämpfungsglied, welche die Phase und Amplitude einer abgegriffenen Kopie des Sendesignals in einer Echtzeit-Rückkopplung von MRT verändern. Ein gradient descent Algorithmus wurde implementiert um die Phasen- und Amplitudenwerte zu finden, die das Sendesignal am besten unterdrücken. Verbleibendes Sendesignal wurde dann während der Rekonstruktion subtrahiert, wozu Referenzmessungen mit einer kleinen Ringspule (Schnüffelspule) verwendet wurden. Der automatisierte analoge Unterdrückungsschaltkreis wurde an einem klinischen 3T MR System getestet, und CEA MRT Aufnahmen eines Phantoms und einer menschlichen Hand wurde akquiriert um die Durchführbarkeit des Konzeptes zu demonstrieren.

Abschließend wurde ein neues System für die Untersuchung des Vokaltraktes entwickelt, welches gleichzeitig mit einer dynamischen MRT Messung der Lunge verwendet werden kann, so dass Grundlagenexperimente zur Lungenphysiologie während des Gesangs durchgeführt werden können. Um die Oszillationsmuster der Stimmlippen während einer dynamischen MRT Messung analysieren zu können, wurde ein Elektroglottographie-(EGG)-system so modifiziert, dass eine gleichzeitige EGG- und MRT-Aufnahme möglich wird. Ein Tiefpassfilter wurde eingefügt um eine HF-Einkopplung in das EGG-Signal zu unterdrücken. Um die MRT-Sicherheit sicherzustellen wurde die HF-Erhitzung in einem Gelphantom getestet, und die Funktionsfähigkeit des Gerätes wurde in einem Probandenexperiment bei Gesangsfrequenzen zwischen A5 (*a´´*) und A3 (*a*) untersucht. In den aufgenommenen EGG-Signalen wurden verbleibende HF-Interferenzen durch ICA-Nachverarbeitung eliminiert, und Standard-EGG-Parameter wie die Fundamentalfrequenz, der Kontaktquotient und der Jitter wurden berechnet. In einem weiteren Probandenexperiment wurden die EGG-Daten mit der Ausdehnung der Lunge verglichen, die aus zeitaufgelösten 2D Messungen gewonnen wurde. Die Messungen zeigten, dass EGG-Messungen an professionellen Sängern während dynamischer Lungen-MRT möglich sind, und dass die Kehlkopfaktivität sicher bei 1.5 T gemessen werden kann.

Nomenclature

Abbreviations

MR	Magnetic Resonance
MRI	Magnetic Resonance Imaging
NMR	Nuclear Magnetic Resonance
CW	Continuous wave
CEA	Concurrent Excitation and Acquisition
IBFD	In Band Full Duplex
EPR	Electron Paramagnetic Resonance
SWIFT	Sweep Imaging with Fourier Transform
ZTE	Zero Echo Time
UTE	Ultra-short Echo Time
SPI	Single Point Imaging
PETRA	Pointwise Encoding Time Reduction with Radial Acquisition
FM	Frequency Modulated
HS	Hyperbolic secant
Tx	Transmit
Rx	Receive
e.m.f.	Electromotive Force
RF	Radio Frequency

CT	Computed Tomography
FID	Free Induction Decay
SE	Spin Echo
CPMG	Carr-Purcell-Meiboom-Gill
FSE	Fast Spin Echo
TSE	Turbo Spin Echo
TrueFISP	True Fast Imaging with Steady State Precession
GE	Gradient Echo
GRE	Gradient Recalled Echo
FLASH	Fast Low Angle Shot
TR	Repetition Time
TE	Echo Time
SNR	Signal to Noise Ratio
ADC	Analog to Digital Converter
DAC	Digital to Analog Converter
LNA	Low Noise Amplifier
RFPA	Radio Frequency Power Amplifier
ULNA	Ultra-low Noise Amplifier
DC	Direct Current
NF	Noise Figure
NPR	Noise Power Ratio
LPF	Low-pass Filter
Q	Quality Factor

PA	Phase and Amplitude
PUC	Pick-up Coil
FOV	Field of View
ROI	Region of Interest
PSF	Point Spread Function
D	Dimension(al)
BALUN	Balanced-Unbalanced Transformer
ESR	Effective Series Resistance
AFI	Actual Flip angle Imaging
BS	Bloch Siegert
HAP	Hydroxyapatite
EGG	Electroglottography
ECG	Electrocardiography
EEG	Electroencephalography
EOG	Electrooculography

Symbols

I_z	Spin number	-
ω_L	Larmor frequency	rad/s
B_0	Static magnetic field in z direction	T
γ	Gyromagnetic ratio	Hz/T
μ	Magnetic moment	A m^2
E	Spin Energy	J
$\hbar$	Reduced Planck's constant	J s
K	Boltzmann constant	J/K
T	Temperature	K
M_0	Net magnetization	A/m
M_{ss}	Steady-state magnetization	A/m
M_{xy}	Transverse magnetization	A/m
M_z	Longitudinal magnetization	A/m
B_1	Time varying excitation RF field	T
α	Flip angle	°
T_1	Spin-lattice relaxation time	s
T_2	Spin-spin relaxation time	s
T_2^*	T_2 - star relaxation time	s
χ	Susceptibility	-
σ	Time domain root mean square noise e.m.f.	A Ω Hz
Δf	Receiver frequency bandwidth	Hz

R_S	Sample resistance	Ω
R_C	Coil resistance	Ω
R_D	Diode series resistance	Ω
L_C	Coil inductance	H
C_{tune}	Tuning capacitor	F
C_{match}	Matching capacitor	F
G	Gradient field	mT/m
Z_0	Characteristic impedance	-
W	RF energy stored in resonator	J
P	Energy lost over one period	J/s
Q_U	Unloaded quality factor of the resonator	-
Q_L	Loaded quality factor of the resonator	-
NF	Noise figure	dB
F	Noise factor	-
V_S	Sample volume	m^3
λ	Electromagnetic propagation wavelength	m
η	Sample's coil filling ratio	-
I	Current	A
U	Voltage	V
T_S	Sampling time	s
Γ	Reflection coefficient	-
Φ	Phase	Rad
$s(t)$	Acquired MR signal	V

$S(\omega)$	Fourier transform of the MR signal	V
$h(t)$	FID response function	-
$H(\omega)$	Fourier transform of the FID response	-
$b(t)$	RF excitation signal function	V
$B(\omega)$	Fourier transform of RF excitation signal	V
$A(t)$	Time dependent leakage signal	V
f_s	Sweep frequency	Hz
TR_{eff}	Effective repetition time	s
t_{acq}	ADC acquisition time	s

Contents

1 Introduction

This thesis describes the novel technologies I developed for structural and functional MR imaging of tissues with ultra-short T_2 values. The technologies were published in several peer-reviewed publications. Additional technical details which are not necessarily novel but help to understand the concepts can be found in the Appendices.

This chapter will start with a historical introduction, because the concept of concurrent excitation and acquisition which forms a major part of this thesis is best understood in a historical context. The historical description follows the thought flow in the textbooks of Fukushima and Cowan, and Nishimura is also instructive to understand all concepts.

1.1 Magnetic Resonance Imaging: A Historical Overview

In the second half of the 19th and the early 20th century many studies were performed to understand the interaction of electromagnetic radiation with matter. For example, the attenuation of X-rays in tissue was first described by Röntgen in 1895, and the heating of water-containing substances in microwave ovens are well-known phenomena of electromagnetic interaction. Typical frequencies for X-rays are on the order of 10^{18} Hz, while for microwave ovens it is around 1-3 GHz.

Magnetic Resonance Imaging (MRI) operates mostly in the very high frequency (VHF) band of the spectrum covering from 10^5 - 0.5 10^9 Hz. The major difference of MRI from other electromagnetic radiation-based techniques is that in addition to the high frequency field, B_1, an external static magnetic field, B_0, is applied. And the second distinguishing property is that the magnetic field component of the incident radiation induces resonant transitions of the nuclei from one quantum mechanical state to another [1]. Thus, the acquired signal is smaller than an electric field interaction by a factor of the velocity in the system or the speed of light.

In the year 1937 Rabi demonstrated that Lithium nuclei can be induced to flip their magnetic moments when exposed to an oscillating magnetic field [2]. Rabi used an electromagnet of B_0 = 0.2 T, and an RF resonator radiating an oscillatory RF field at 3.5 MHz. He varied the static magnetic field strength by adjusting the current in the electromagnet, and then let LiCl molecules pass through a vacuum chamber into the magnetic field. He could observe resonance peaks at certain field strengths due to absorption of the continuously radiated RF energy – he named this phenomenon nuclear magnetic resonance (NMR).

In 1945, Purcell filled an electromagnetic cavity with solid paraffin and placed it in an electromagnet. An RF wave at 30 MHz was continuously applied to the cavity, and the magnetic field strength was slowly increased until at about 0.7 T. At resonance the cavity's absorption of RF radiation energy increased by 20 fold [3] - this was the first demonstration of NMR of hydrogen nuclei.

At the same year, Bloch demonstrated NMR in water using a much smaller apparatus. A small glass tube was filled with 1 cc of water. Separate transmitter and receiver resonators were tuned at 8 MHz, which were placed orthogonal to each other for isolation i.e. geometrical decoupling. An electromagnet operating at approximately 0.18T was used and, similar to Purcell, Bloch team slowly changed the magnetic field until resonance was achieved. Rather than measuring absorption, however, Bloch detected a nuclear induction signal in the receiver coil as a manifestation of the NMR phenomenon [4].

After these early observations several researchers, most notably Erwin Hahn in 1950, demonstrated that similar information could be gathered by observing the response to short intense pulses of resonant RF radiation [5].

Thus, the methods for MR signal acquisition can be divided historically into two major classes:

1. Methods, that excite and acquire signals simultaneously, which will be termed Concurrent Excitation and Acquisition (CEA) methods in this work such as Continuous wave NMR (CW-NMR), rapid scan correlation spectroscopy, stochastic NMR, and
2. Time interleaved acquisition methods in which excitation and acquisition are not performed simultaneously, which will be called pulsed NMR such as spin echo, gradient echo, steady-state free precession, ultra-short TE, or single point imaging.

CEA was historically the first MR technique, which, at that time, was called continuous-wave (CW) nuclear magnetic resonance (NMR). The discovery of NMR by Bloch and Purcell, and the first MR image by Lauterbur [6] were also based on CW-NMR. In CW-NMR either a frequency-swept RF pulse was used or the static field was swept through resonance, and it has been used extensively to spectroscopically investigate solid materials with ultra-short T_2*[7], and has been adapted to ^1H and multinuclear imaging to reduce the dead time between RF excitation and signal acquisition. In 1974, it was shown that the same CW absorption spectra could also be acquired with a rapidly swept RF pulse (i.e., linear region of an adiabatic passage) that correlates with the spin response [8], [9] – this rapid scan correlation method was further improved for 3D solid MRI [10]–[13].

In 1975, the MRI based on time interleaved acquisition [14] was successfully demonstrated, and since then, pulsed NMR methods have become the most popular techniques in MRI, since they are easier to implement and they are often sufficient for clinical imaging purposes. The development of MRI electronics and the increase in computational power also led to a domination of pulsed NMR techniques in MRI.

1.2 Motivation and Main Contributions

MRI produces images of tissue and materials noninvasively using non-ionizing RF waves, and it provides information about tissue structure, metabolism, and function. The impact of MRI in diagnostic radiology is very high, and, despite its wide-range clinical and commercial use, MRI is still an active area of research. Currently, new application areas are opened through the development of high-field MR systems at 7 Tesla and higher, and new encoding methods that overcome the limitations of linear gradient systems.

Most MRI examinations between the years 2011-2015 are applied in Turkey, USA, and Germany, and the numbers tend to monotonically increase every year [15]. This motivates efforts to make MRI more efficient (i.e., faster and with improved image quality) and more comfortable. Presumably, the increase in MRI usage is partly stimulated by the research on new contrast mechanisms [16], [17], real-time imaging applications [18], expanding usage of MRI to the patients with implants [19], and investigating brain function and metabolism [20], [21].

Nevertheless, MRI still has several fundamental limitations:

- long scan durations
- artifacts in images caused by subject motion
- very high acoustic noise levels
- signal loss due to short relaxation times
- RF induced heating of electrically conducting objects

For patients an MRI exam is often uncomfortable as they have to lie still in the magnet bore for tens of minutes trying to freeze all the movements, they are covered with RF resonators, and they are exposed to the loud noise. Unlike computed tomography or x-ray scans, image encoding in MRI takes longer time as the spatial information is collected line by line as on a transformed information space in frequency domain.

Currently, significant research effort is focused on reducing the MRI scan time: parallel MRI using multi-element receive antennas [22], rapid acquisition techniques manipulating the spin echo signal [23], [24], compressed sensing applications in MRI [25], and echo planar imaging (EPI) methods. The advantages, disadvantages, and limitations of these technologies will not be discussed in this work, because the proposed CEA method can be combined with all of them, and constitutes an additional method to accelerate the image acquisition. CEA approaches the problem from a different perspective: instead of separating transmit and receive events in time, CEA is a technology that enables concurrent transmission and reception to reduce the scan times in MRI. This technology will be described in detail under "Concurrent Excitation and Acquisition". Data acquisition strategies using CEA have additional advantages such as silent scanning, and reduced sensitivity to motion artifacts due to radial frequency encoding.

In several tissues the MR signal decays rapidly due to ultra-short T_2, so that signal from these tissues is not detectable using conventional MRI techniques. Until the 1990s, these tissues were considered 'undetectable' with MRI [26]. At this time, various novel data acquisition methods were suggested [27]–[29], and modifications to these are still an active field of research. Common to all these methods is to apply a very short 'hard' excitation radio frequency (RF) pulse, and to acquire the FID signal – rather than echoes – immediately after this pulse to shorten the delay between signal generation and data acquisition. Major limitations of these ultra-short TE methods are caused by the used RF hardware: most RF power amplifiers do not support very short and high peak RF pulses, a certain amount of time is needed to switch between the RF transmit and receive mode, and high-order digital filters put a lower limit on the time between excitation and acquisition, i.e. acquisition delay. In this thesis, technologies are developed to implement state-of-the-art acquisition methods to MRI of samples with extremely short T_2 values (e.g. ancient mummified tissue), and, for the first time, parametric relaxation time mapping of such samples is done using MRI. However, CEA with its intrinsic property of having zero acquisition delay might be the ultimate solution to MRI of samples with ultra-short T_2.

Lungs have a very short T_2 value, and the study of lung motion in combination of vocal fold activity can provide deep insights into breathing strategies of patients with lung function abnormalities or singers trying to obtain a high vocal quality. In this thesis we describe a novel technology that enables simultaneous measurement of dynamic lung MRI and vocal cord activity with a so-called electro-glottography (EGG) system. One problem of MRI with external devices such as the EGG is RF-induced heating of electrically conducting elements inside the RF field: with more than 150 occasions reported, RF heating can result in severe burns in patients due to metallic electrodes or implants [30]. As in microwave heating, RF electric field coupled in the metallic structures such as implants, pacemaker leads, interventional instruments, electrophysiological measurement electrodes are potentially harmful to the patients and volunteers in MRI exams. A great research effort is spent on avoiding harmful effects of RF induced heating in MRI. Electrophysiological measurements, however, are significant in MRI exams to track patient status, to record functional information, and to apply triggered imaging techniques of moving organs such as the heart. Most electrophysiological measurement devices are tested, and for example ECG electrodes exist that are certified as 'MR compatible' or 'MR conditional' according to the international standards. EGG electrodes, which are used to monitor functional vocal fold activity, were never introduced to MRI until now. Regarding RF induced heating of metallic implants in MRI, concurrent excitation and acquisition technology is also advantageous as it has very low peak RF excitation powers. This makes the technology potentially promising to offer safe MRI examination to the patients with implants.

1.3 Thesis Scope and Organization

This thesis describes novel MRI technologies and applications, and it is divided into three main topics:

- investigation of state-of-the-art pulse sequences and hardware improvements for MRI of samples with ultra- short transverse relaxation times and application to MRI of an ancient mummified sample,
- novel systems for MRI with Concurrent Excitation and Acquisition, and
- simultaneous measurement of EGG signal and functional imaging of lungs

This thesis is divided into theory, methods, results, and discussion chapters. Each chapter is also separated into sections for each three main topics. Electronic circuit layouts and simulations, derivations of formula, and programming scripts are detailed in designated Appendix chapters. In the following section of this chapter, a description of the phenomenon of nuclear magnetism will be given and the invention of MRI will be discussed based on a comparison of CEA and pulsed NMR techniques.

In chapter 2, basic MRI principles will be introduced, and the theoretical framework used throughout the thesis will be developed. Starting with signal acquisition strategies, relaxation mechanisms are described and the motivation for each of the studies is stated. State-of-the–art pulse sequences for MRI of samples with ultra-short T_2 including UTE, SPI, and PETRA will be described, as well as contributions to improvement of these techniques. Hardware considerations will also be discussed. Novel techniques to implement CEA in clinical MRI systems will be described. Finally background for a novel system that enables simultaneous EGG recordings and dynamic lung MRI measurements will be established.

In chapter 3, methods for each main topic is described such as technical details and experimental setups. Application of the state-of-the-art pulse sequences to MRI of an ancient mummified human hand as an extreme case of the samples with ultra-short T_2 will be presented. Modifications to the existing pulse sequences and optimization of hardware will be described. Novel systems for decoupling of Tx and Rx resonators for MRI with CEA will be explained.

The CEA section is divided into two parts: first part describes the active decoupling method based on destructive interference of multiple Tx resonators. In this part of the study, CEA is implemented by combining geometric decoupling with a novel method that adjusts the relative phase between two separate Tx resonators. This method attempts to cancel the B_1-induced currents on the Rx resonator by an appropriate adjustment of the amplitudes and phases of the Tx resonator array input currents. The method was tested for CEA imaging of short-T_2* phantoms in a 3T clinical MRI system.

The second part of the CEA section describes the latest developments in my research. In this section, an automated analog cancellation system is described which is designed to bring CEA one step closer to clinical practice. The decoupling system designed in this work benefits from all decoupling methods used in full duplex electronics: passive suppression using geometrical decoupling, active cancellation using an analog cancellation circuit, and digital cancellation. Additionally, an automatized decoupling procedure is presented using real-time feedback control between the analog cancellation circuit and the MR system. Finally the system for simultaneous EGG and dynamic lung MRI will be described along with the MRI safety issues.

The performance results of the systems parts from the previous section are given in chapter 4. Analysis of the data for parametric imaging of ancient mummified sample are presented, which are the first parametric maps for ancient mummies using MRI. The first proof-of-principle results including in vivo MRI with CEA will be analyzed. For the simultaneous EGG and MRI system, results of the safety tests will be given, and simultaneously acquired vocal fold activity and dynamic lung MRI measurement results during a singing task of a trained singer are shown.

In chapter 5, the results for each main topic will be discussed. Limitations and future developments will be highlighted. Potential value of the systems for MRI in general will be evaluated.

2 Theory

2.1 Basic NMR Principles

An atomic nucleus has four properties: mass, electric charge, magnetism, and spin. The spin property of a nucleus indicates that it rotates in space. A rotating object possesses angular momentum. Spin is also a form of angular momentum. Microscopic magnetic field generated by spin is represented with a magnetic moment vector, μ. Ratio of magnetic moment and spin angular momentum is called gyromagnetic ratio, γ, which is an intrinsic property of the nucleus. The nuclear magnetic resonance (NMR) phenomenon is caused by the magnetic moment of the atomic nucleus [1]. When exposed to an external magnetic field, B_0 all nuclei with a non-zero magnetic moment precess about the field axis. The precession frequency, also called Larmor frequency, ω_L is proportional to the strength of the external magnetic field, B_0, and γ:

$$\omega_L = -\gamma |B_0| \tag{2.1.1}$$

The hydrogen nucleus (proton, ^{1}H) is the most sensitive stable NMR nucleus with $\gamma = 2\pi$ 47,577 MHz/T (i.e., a resonance frequency of 128 MHz at B_0 = 3 T).

Nuclear spins are associated with a quantum number, I_z. ^{1}H isotope of hydrogen has $I_z = ½$, implying that magnetic moment can be parallel or anti-parallel with B_0. ^{2}H isotope, however has $I_z = 1$, when proton and neutron combined in parallel configuration, and $I_z = 0$ in anti-parallel configuration. Since the energy level difference between these two states exceeds the energies available to usual electromagnetic fields, excited states of such nuclei is ignored. The nuclear magnetic moment is a consequence of the nuclear spin state, I_z, and can be written as $\mu = -\gamma\hbar I_z$. The interaction energy of a magnetic moment μ in a magnetic field B_0 is μB_0. Thus, the energy, E is given by

$$E = \mu B_0 = -\gamma \hbar B_0 I_z \tag{2.1.2}$$

Here, it was assumed (without loss of generality) that B_0 is parallel to the z-axis. This form of magnetic interaction is known as a Zeeman interaction. The two values of the spin quantum number I_z for ^1H atom are –½ to +½. The difference between the spin state energy levels writes

$$\Delta E = E(I_z = -½) - E(I_z = +½) = \gamma \hbar B_0 \tag{2.1.3}$$

So far, the quantum mechanical description explains the interaction of a single nuclear spin with a magnetic field. In the following, the observation will be extended to a macroscopic spin ensemble.

In an MRI experiment a large ensemble of nuclei is observed, where different nuclei may occupy different states. Therefore, an ensemble average for the individual magnetic moments and spin states needs to be considered. In a proton ensemble, two groups of nuclei states, one (n_+) parallel to B_0, and the other (n_-) anti-parallel. Since the n_+ group represents the lower energy state, it would be expected that many more spins occupy this state than n_-. However, as the thermal energy $E_{thermal} = kT$ is at room temperature much larger than the magnetic energy ΔE (high temperature approximation), the probability ratio of two spin states is described by a Boltzmann distribution:

$$\frac{n_+}{n_-} = e^{\Delta E/kT} \tag{2.1.4}$$

Here, k is Boltzmann's constant and T is the absolute temperature. To calculate the equilibrium nuclear magnetization we need to determine the difference between two populations with opposite polarity, and the average spin magnetic dipole energy. First, we approximate eqn.2.1.4 using Taylor series expansion:

$$\frac{n_+}{n_-} \approx 1 - \Delta E/kT \tag{2.1.5}$$

Thus,

$$\Delta n = n^+ - n^- = n^+ \, \Delta E/kT \sim (\Delta E/kT) n_s/2 \tag{2.1.6}$$

where, n_s is the concentration of the nuclei in the body. The resulting equilibrium magnetization for ^{1}H spins can be calculated as

$$M_0 = \Delta n \hbar \frac{\gamma}{2} = n_s \gamma^2 \hbar^2 B_0/4kT \tag{2.1.7}$$

More generally M_0 is dependent on the spin state [31]:

$$M_0 = n_s \gamma^2 \hbar^2 I_z (I_z + 1) B_0/3kT \tag{2.1.8}$$

This equation shows that the magnetization is proportional to the B_0 field strength. At B_0 = 3 T, T = 300°K, magnitude of the magnetization vector for ^{1}H protons in H_2O corresponds to 9.75 10^{-2} A T/m.

Time dependent behavior of the ensemble of a spin system can be modeled using a bulk magnetization vector $\boldsymbol{M}$, which is the sum of a large number of individual nuclear magnetic moments.

2.1.1 Bloch Equations

Equation of motion for $\boldsymbol{M}(t)$ is modeled as a forced mechanical rotation, where the torque is equal to the rate of change of $\boldsymbol{M}(t)$ for magnetic field vector $\boldsymbol{B}(t)$:

$$\frac{d\boldsymbol{M}(\boldsymbol{t})}{dt} = \gamma \boldsymbol{M}(\boldsymbol{t}) \times \boldsymbol{B}(\boldsymbol{t}) \tag{2.1.9}$$

giving the so called Bloch equations [32]. Suppose that $\boldsymbol{B}(t)$ = [0, 0, B_0]'. If initial magnetization $\boldsymbol{M}(t)$ were oriented at an angle α relative to the z-axis, the solution of Eq. 2.1.9 is

$$M_x(t) = M_0 \sin\alpha \cos(-\omega_L t + \phi)$$

$$M_y(t) = M_0 \sin\alpha \sin(-\omega_L t + \phi)$$

$$M_z(t) = M_0 \cos\alpha$$

where $M_0 = |M(0)|$, $M(t) = [M_x(t), M_y(t), M_z(t)]$, and ϕ is an arbitrary angle. These equations describe a precession of $M(t)$ around B_0 with a frequency $\omega_L = \gamma B_0$.

For a better understanding of MRI, it is useful to conceptualize $M(t)$ in two components: longitudinal magnetization is oriented along z-axis, and transverse magnetization is defined by $M_{xy}(t)=M_x(t)+iM_y(t)$. M_{xy} is a complex variable, angle of which is called the phase angle and given by

$$\phi = \tan^{-1}\frac{M_y}{M_x} \tag{2.1.10}$$

Transverse magnetization can be written as

$$M_{xy}(t) = M_0 \sin\alpha\, e^{-i(\omega_L t - \phi)} \tag{2.1.11}$$

In MRI theory, evolution of the magnetization is expressed in the rotating frame, which is rotating at ω_L, for convenience. The coordinates of the rotating frame are related to those of stationary frame by

$$x' = x\cos(\omega_L t) - y\cos(\omega_L t)$$

$$y' = x\sin(\omega_L t) - y\cos(\omega_L t)$$

$$z' = z$$

In this frame of reference, transverse magnetization becomes

$$M_{xy}(t) = M_0 \sin\alpha\, e^{i\phi} \tag{2.1.12}$$

In other words, M_{xy} is a stationary vector in the rotating complex plane.

2.1.2 RF Excitation and Flip Angle

It is possible to tip $\boldsymbol{M}(t)$ away from B_0 using an oscillating magnetic field, $[B_1(\mathrm{t}), 0, 0]'$, which has a frequency ω_{RF}.

$$B_1(t) = |B_1(t)|e^{-i(\omega_{RF}t-\varphi)} \tag{2.1.13}$$

In the rotating frame of reference:

$$B_1(t) = |B_1(t)|e^{-i(\omega_{RF}-\omega_L)t-\varphi)} \tag{2.1.14}$$

Assume that $\omega_{RF} = \omega_L$, and for simplicity initial phase, $\varphi = 0$; so that B_1 is oriented in x direction of the rotating frame (Fig.1). Such an excitation causes $\boldsymbol{M}$ to precess in the clockwise direction in the yz-plane. The frequency of this precession is given by $\omega_1 = \gamma|B_1|$. Final tip angle, i.e. flip angle, α, and phase of the $\boldsymbol{M}$ depends both on the amplitude and duration of $B_1(t)$. If the RF excitation field is turned off after $\boldsymbol{M}$ has precessed into the transverse plane, then pulse is called $\pi/2$ pulse. The RF excitation that tips $\boldsymbol{M}$ to $\boldsymbol{-M}$ is called π or inversion pulse. In general,

$$\alpha = \gamma \int_0^{T_{RF}} B_1(t)dt \tag{2.1.15}$$

In the case of $\omega_{RF} \neq \omega_L$, as $\boldsymbol{M}$ is tipped away from the z-axis, B_1 moves relative to the $\boldsymbol{M}$, and the axis of rotation is changed. As a result, there is a residual longitudinal magnetization component in z-axis. Difference between ω_{RF} and ω_L, $\Delta\omega = \gamma B_z$ characterizes the longitudinal component of $\boldsymbol{M}$. Together with B_1, B_z produce an effective B_1 field, B_{eff}, in the rotating frame. When B_1 is applied off-resonance (i.e., slightly different from the Larmor frequency), $\boldsymbol{M}$ precesses in a cone around B_{eff} at frequency $\Delta\omega$ instead of precessesing around B_1 (Fig.1c). If $|\Delta\omega| < \omega_1$, then effective excitation occurs; if $|\Delta\omega| \gg \omega_1$, then no excitation occurs.

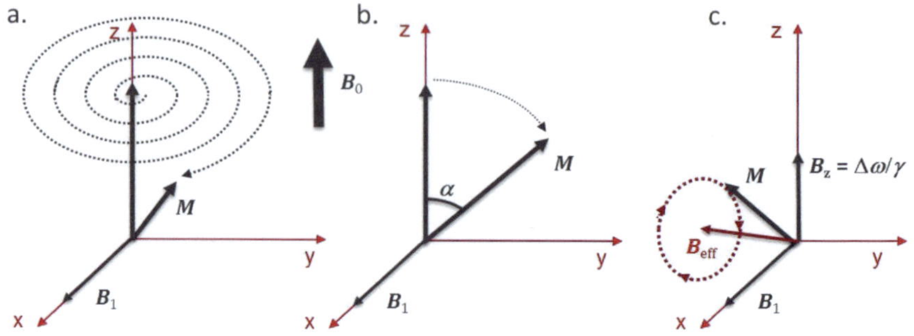

Fig. 1: RF Excitation and magnetization vectors. In the lab frame, actual motion of M is a spiral (left). In the rotating frame M precesses around the x-axis in response to an RF excitation, B_1 (middle). When B_1 is off-resonant with respect to Larmor frequency, M precesses in a cone around B_{eff} at frequency $\boldsymbol{\Delta\omega}$.

Including B_1(t), Bloch equations are written in matrix form as [33]:

$$\frac{d\boldsymbol{M(t)}}{dt} = \gamma\boldsymbol{M(t)} \times \boldsymbol{B_{eff}(t)}$$

$$\frac{d\boldsymbol{M(t)}}{dt} = \begin{pmatrix} 0 & \omega_L - \omega_{RF} & 0 \\ \omega_{RF} - \omega_L & 0 & \omega_1(t) \\ 0 & -\omega_1(t) & 0 \end{pmatrix} \boldsymbol{M(t)} \tag{2.1.16}$$

The solution of Eq. 2.1.16 is derived as:

$$\boldsymbol{M(t)} = \begin{pmatrix} 1 & 0 & 0 \\ 0 & \cos\left(\int_0^t \omega_1(\tau)\mathrm{d}\tau\right) & \sin\left(\int_0^t \omega_1(\tau)\mathrm{d}\tau\right) \\ 0 & \sin\left(\int_0^t \omega_1(\tau)\mathrm{d}\tau\right) & \cos\left(\int_0^t \omega_1(\tau)\mathrm{d}\tau\right) \end{pmatrix} \boldsymbol{M(0)} \tag{2.1.17}$$

corresponding to a rotation about x-axis, the direction of the applied $B_1(t)$.

In MRI two types of RF excitation is applied: non-selective and selective.

2.1.2.1 Selective Excitation

B_1 is generated by a resonator, which is an electrical device that converts electric current into electromagnetic waves (please refer to the section 2.2 for detailed description of resonators). In the presence of only B_0, B_1 excites all spins in the volume sensitive to the magnetic field generated by the resonator. If $B_1(t)$ is applied in presence of a linear static field gradient along z, G_z, then only those spins residing in z-locations with resonant frequencies matching the frequencies of $B_1(t)$ will be excited [34]. Spins residing in z-locations with resonant frequencies outside the bandwidth of $B_1(t)$ will remain unexcited. In this case, excited plane will be perpendicular to z-axis (Fig.2).

In selective excitation, slice profile can be approximated by the Fourier transform of the B_1(t). For small tip angles (α < 30°) Bloch equations can be linearized, which results in following solution for the time evolution of the transverse magnetization:

$$M_{xy}(z,t) = iM_0 e^{-i\omega(z)t} \int_0^t e^{i\omega(z)\tau} \omega_1(\tau) d\tau \qquad (2.1.18)$$

This can be interpreted as a Fourier transform as shown in eqn.2.1.19 for the general case of a time varying gradient *G*(t) [33]. The frequency domain of the transformation is called the *excitation k-space* and characterizes the profile of the selected slice.

$$M_{xy}(t,\vec{r}) = iM_0 \int_0^t \omega_1(\tau) e^{-i2\pi\vec{k}(\tau)\cdot\vec{r}} d\tau \qquad (2.1.19)$$

for

$$\vec{k}(\tau) = \frac{\gamma}{2\pi} \int_\tau^t \vec{G}(t')dt' \qquad (2.1.20)$$

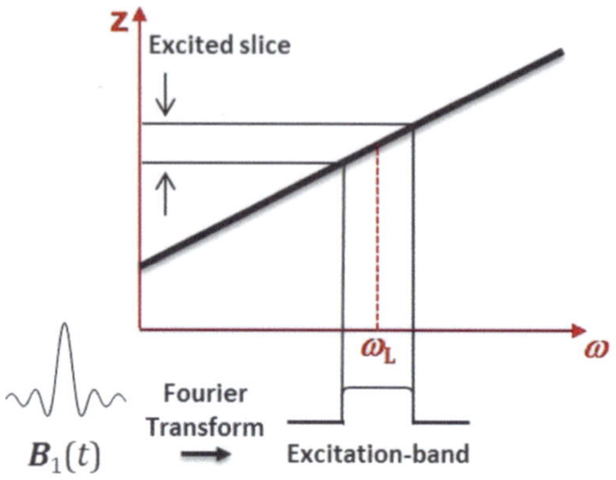

Fig. 2: Selective excitation. Fourier transform of $B_1(t)$ approximates the slice profile.

Precession behavior and tilting of the magnetization using RF excitation are not sufficient to explain the behavior of magnetization. Since the spin behavior is a thermodynamic process, the acquired energy from the RF pulse is released after the RF pulse is switched off, and spins return to the equilibrium state.

2.1.3 Relaxation Mechanisms

2.1.3.1 Spin-lattice Relaxation: T_1

In equilibrium state, the magnetization vector has only a longitudinal component along z-axis while the transverse component is zero (Fig. 3). After application of an α pulse ($\alpha \neq \pi$), relaxation mechanisms dampen the otherwise constant precession movement of $\boldsymbol{M}$, and bring it back to the equilibrium. After a spin state change due to an external energy transfer, spin groups change to their new equilibrium value by an exponential process, where the time constant for this transmission is called the *spin-lattice* relaxation time, T_1 [32]. The term *lattice* refers to the other nuclei and electrons surrounding the spin. The shorter T_1, the faster the equilibrium state is reached, i.e. the spin, as a magnetic dipole, has a more effective coupling with the lattice. The range of T_1 values in human tissue is between 10^{-1} to 2 s [35]. T_1 can be as high as 10^2 s in solids, which caused the first NMR experiments to fail as the signal saturated before it could be detected [36].

2.1.3.2 Spin-spin Relaxation: T_2

Spins are magnetic dipoles and generate their own local field. As a result of the coupling between dipolar fields of two adjacent spins with opposite polarity, they can exchange energy. Such a *spin-spin* interaction does not affect the total system energy but the life time of the excited state. The combination of B_0 and the local dipolar field is different for the spins at different locations. As a result, instead of one resonance condition, there are many spins precessing at varying frequencies. If we consider a signal induced by many out-of-phase signal sources, the stronger the spin-spin interactions the faster the induced signal would decay due to de-phasing of the spins. This is called T_2 relaxation [37].

In the free induction decay (FID), an apparent transverse relaxation T_2* is observed that is shorter than T_2, which is caused by the inhomogeneity of B_0 due to instrumental imperfections:

$$\frac{1}{T_2^*} = \frac{1}{T_2} + \frac{1}{T_2'} \qquad (2.1.21)$$

T_2 includes variations due to molecular and nuclear phenomena, and T_2' is caused by field variations (susceptibility changes) on the microscopy, mesoscopic and macroscopic scale. Detailed explanation on the sources of strong T_2* effects in human tissue will be given in section 2.3.

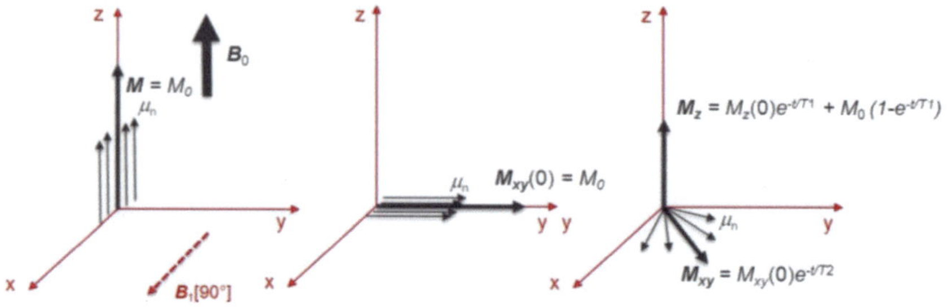

Fig. 3: Spin behavior demonstrated for the time frames of equilibrium (left), after excitation of 90° flip angle (middle), longitudinal (T_1) and transverse (T_2) relaxation (right). After the spins lose all their energy and become completely dephased, equilibrium state is reached again.

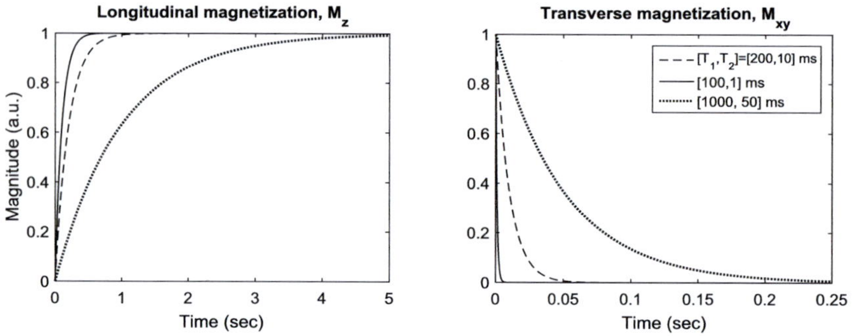

Fig. 4: Magnitude of longitudinal (M_z) and transverse (M_{xy}) magnetization during relaxation for various relaxation parameters. The relaxation of magnetization parallel to the B_0 field: the approach towards M_0 and the decay of magnetization in the transverse plane to zero. The spin-lattice relaxation time, conventionally denoted by T_1 characterizes the relaxation of magnetization parallel to B_0, while the spin-spin relaxation time, T_2, characterizes the relaxation in the plane perpendicular to the B_0 field.

The relaxation behavior of $\boldsymbol{M}$ is summarized in Fig. 3. After a $\pi/2$ pulse tilts the total magnetization vector, energy transitions due to *spin-lattice* interaction affect the longitudinal component of the magnetization, $\boldsymbol{M_z}$, while *spin-spin* interactions and magnetic field inhomogeneity cause the transverse magnetization, $\boldsymbol{M_{xy}}$ to reach an equilibrium state, which is zero. In Fig. 4, magnetization components during relaxation is given calculated for various spin groups with (T_1,T_2) = (200 ms, 10 ms), (100 ms, 1 ms), and (1000 ms, 50 ms).

Relaxation terms are added to the Bloch equations in rotating frame:

$$\frac{d\boldsymbol{M}(\boldsymbol{t})}{dt} = \gamma\boldsymbol{M}(\boldsymbol{t}) \times \boldsymbol{B}(\boldsymbol{t}) - \frac{\boldsymbol{M}_x(\boldsymbol{t}) + \boldsymbol{M}_y(\boldsymbol{t})}{T_2} - \frac{\boldsymbol{M}_z(\boldsymbol{t}) - \boldsymbol{M}_0(\boldsymbol{0})}{T_1} \qquad (2.1.22)$$

This equation evolves such that M_x and M_y relax to zero, while M_z relaxes to M_0. A general analytical solution was also presented in 1994 which includes the behavior of the magnetization vector during RF excitation [38].

2.1.4 MR Signal Generation

To observe the NMR phenomenon, the magnetization vector must be transformed into a measureable quantity. The transverse magnetization given in Eq. 2.1.11 describes a rapidly rotating magnetic dipole, which is the source of the NMR signal measured as a voltage induced in a resonator (cf. section 2.2). Transformation from magnetic to electric domain is realized by Faraday's induction principle. In fact, Faraday's induction is used both to manipulate nuclear spins (i.e., RF excitation) and to generate signals from active spins.

Faraday's law of induction states that time-varying magnetic field cutting across a coil of wire will induce a voltage in the wire, which is referred to as *electromotive force (e.m.f.)*. The term 'coil of wire' refers to the resonator, which is a special type of detector that is tuned at the Larmor frequency in order to have the most sensitivity at the frequency band of interest. The resonator is oriented such that B_1 is perpendicular to B_0 and applies a transverse magnetic field. Spins are excited out of equilibrium and the resonator detects changes of magnetization in the xy-plane, i.e. transverse magnetization (Fig. 5). The simplest NMR experiment consists of a single uniform RF excitation pulse (rectangular pulse), after which the spins of a sample are excited and induce a *free induction decay (FID)* signal. This pulsed experiment is, for example, applied by the conventional MRI systems to detect the resonance frequency of the sample and to determine the RF pulse amplitude necessary to produce the maximum signal.

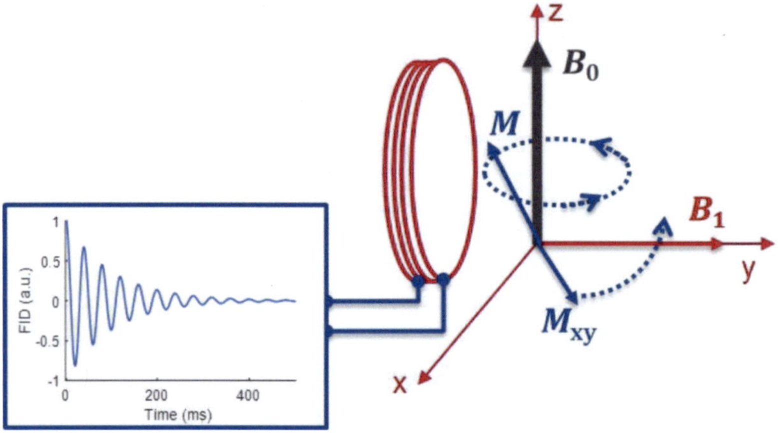

Fig. 5: Behavior of magnetization after excitation. Transversal component of the precessing magnetization induces a measureable voltage signal in a resonator. The generated signal is called FID, which is a damped sinusoid due to combined of precession and relaxation effects.

Lenz's law dictates that the magnitude of the transverse magnetization is proportional to the e.m.f. Although Eq. 2.1.4 implies that only 2 10^{-6} spins effectively contribute to the thermal magnetization at room temperature, it is still possible to obtain a detectable signal from tissue due to the high spin density (Avogadro number). The MR signal amplitude is also proportional to B_0 and γ (Eq. 1.7), and as ^{1}H nuclei have a high γ and 100% natural abundance, they are the optimal nuclei to acquire a high MR signal [39].

The role of the receiver in the signal acquisition process was explained by Hoult in [40], where the principle of reciprocity was employed to show that the signal induced by a magnetic dipole M within a volume element dV_s is

$$S = -\frac{d}{dt}\int \frac{B_1}{i} M dV_s \qquad (2.1.23)$$

giving the in phase signal component:

$$S = \omega_L \aleph (\tfrac{B_1}{i})_{xy} M_0 V_s \cos(\omega_L t) \tag{2.1.24}$$

$\frac{B_1}{i}$ is the magnetic field at the location of the spins, which is produced by the unit current flowing in the resonator. $(\frac{B_1}{i})_{xy}$ is the transverse component of the magnetic field. $\aleph$ represents the result of integrating over the field.

2.1.5 Detection of the MR Signal

Generation and detection of the MR signal can be achieved in two ways:

1. Pulsed or interleaved methods, where the RF excitation and signal acquisition events are separated in time
2. Concurrent excitation and acquisition (CEA)

To describe both methods the response of the MR system (i.e., the transverse magnetization or the MR signal) is assumed to be linearly related to the input to the system, i.e. the RF pulse $B_1(t)$. This assumption is only correct for small flip angles, as the Bloch equations can only be linearized under this assumption.

2.1.5.1 Interleaved Excitation and Acquisition: From FID to Echoes

If the spins are considered as a linear system, then the resonance spectrum is the frequency response function of the system, and the FID signal is the unit impulse response. The frequency response function, $H(\omega)$, and the unit impulse response, $h(\mathrm{t})$ of a linear system form a Fourier transform pair. Thus, it is possible to obtain all the necessary information about the spin system after a simple Fourier transformation of the FID. This type of signal acquisition was called *'pulsed NMR'*, and the term is generalized as *interleaved excitation and acquisition* to comply with CEA. In 1966, Ernst claimed that measuring FID has a few advantages over CEA (here, slow passage CW NMR, please refer to the section 2.1.5.2.) [41]:

1. The acquisition time to obtain spectra is shorter
2. Sensitivity (i.e., the information per unit time) is higher as all the spins are simultaneously excited
3. Interleaved methods allow investigating nuclear Overhauser effects
4. Frequency calibration of the spectrum is simplified
5. Intermediate passage causes line shifts and deformations (this problem was addressed later in rapid scan correlation methods (cf. section 2.1.5.2)

If FID is recorded with repetitive RF pulses, steady state magnetization is formulated with respect to the repetition time interval, TR >> T_1, and α due to the RF pulse in [41]:

$$M_\alpha = \frac{M_0(1-e^{-TR/T_1})}{1-\cos(\alpha)e^{-TR/T_1}}\sin\alpha \qquad (2.1.25)$$

This equation known as the FLASH equation (cf. section 2.1.5.1.2.).

2.1.5.1.1 Spin Echo

In 1950, Hahn observed in an experiment with more than one RF pulse that an increased signal, the so-called echo, could be observed after the last pulse. He interpreted this signal as a constructive interference of precessing magnetization vectors, and named the phenomenon a *spin echo.* Spin echoes are less sensitive to magnetic field inhomogeneities, and they can be used for example to measure transverse relaxation times [42].

In an inhomogeneous magnetic field the spin system consists of several ensembles which will have different local precession frequencies (isochromat). After a $\pi/2$ pulse, magnetization of each isochromat will precess with a slightly different frequency, and, therefore, the net magnetization will gradually get out of phase resulting in an FID signal with shorter T_2^* decay. If a π pulse is applied after a time interval τ, the magnetization of each isochromat will be flipped by π. Now the accumulated phase differences between the isochromats are reversed as shown in Fig. 6. As the precession proceeds, the dephasing of the ensembles is gradually reversed. When all the ensembles are back in phase (refocusing), the total magnetization reaches at maximum, and a constructive signal superposition, the spin echo (SE) signal is detected by the resonator. The time between π/2 pulse and the echo maximum is called *echo time* (TE). The amplitude of the spin echo decreases with increasing TE, due to random fluctuations in the local fields, which lead to an irreversible decoherence between the spin isochromats. SE detection eliminates the otherwise dominant effects of the non-uniformity of the magnetic field, i.e. the T_2' effects. If the π/2 - π sequence is repeated for several different values of TE, the spin echo signal decreases with e^{-TE/T_2}.

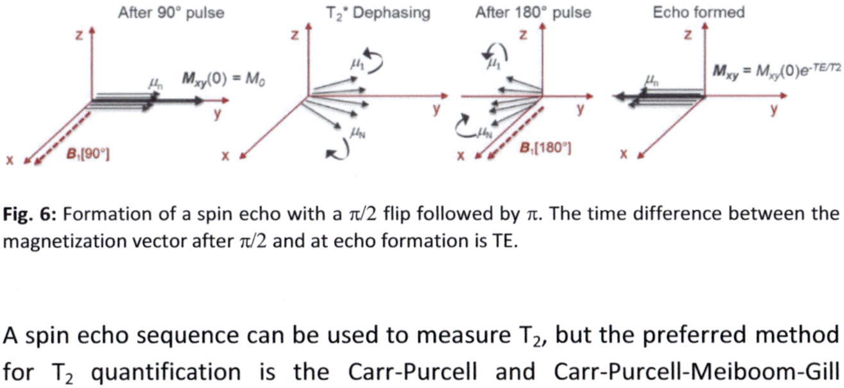

Fig. 6: Formation of a spin echo with a π/2 flip followed by π. The time difference between the magnetization vector after π/2 and at echo formation is TE.

A spin echo sequence can be used to measure T_2, but the preferred method for T_2 quantification is the Carr-Purcell and Carr-Purcell-Meiboom-Gill (CPMG) sequence. The CPMG sequence determines T_2 in a single measurement, and minimizes errors from diffusion. Here, again an initial π/2 pulse is applied, and at times t = TE/2, 3TE/2, 5TE/2, etc. a π pulse is applied, which creates a series of spin echoes at TE. 2TE, 3TE etc. The height of each successive echo is a function of TE.

In 1986, Hennig used the spin echo series to accelerate SE imaging: Rapid Imaging with Refocused Echoes, also known as Fast Spin-Echo (FSE) or Turbo

Spin-Echo (TSE), is a rapid imaging technique that speeds up image acquisition by acquiring a phase encoded line after each π pulse [23]. Combined with the use of crusher or spoiler gradients, this technique has become one of the most commonly used pulse sequences in clinical MRI, as it can accelerate SE MRI by factors of 3-100 yielding a full MR image in down to 200 ms.

2.1.5.1.2 Gradient echo

The phase-sensitive signal formation can be developed also for the spins accumulating phase due to an externally applied gradient field for frequency encoding purposes (cf. section 2.1.5). In 1986, another rapid imaging technique was presented and has become one of the most commonly used pulse sequence in MRI: FLASH (fast low-angle shot) [43]. Here, the initial slice selective RF pulse is not followed by a π refocussing pulse and the different spin isochromats are caused only by the different imaging gradients. The isochromats are refocused if the area under the gradient-time curve vanishes; this is called a gradient echo (GE). The first imaging pulse sequence using GE for data acquisition was called FLASH (fast low angle shot), and it is advantageous over SE due to its lower RF power deposition, and significant reduction of internal waiting times.

2.1.5.2 Concurrent Excitation and Signal Acquisition

Chronologically, concurrent RF excitation and MR signal acquisition was used before the interleaved methods such as FID, SE, and GE were invented. The concept of continuous wave (CW) excitation is that the characteristics of a resonance phenomenon can be analyzed by measuring system's response to a sinusoidal disturbance with varying frequency. In CW NMR, frequency response of the system to a weak oscillating magnetic field is measured directly, whereas in pulsed NMR or time-interleaved methods, a strong RF pulse is used to create a non-equilibrium state, and the time response is measured.

In CW NMR, the spin response can be detected by measuring the absorbed RF power due to the interaction of the RF magnetic field with the spins (similar to a quality factor (Q) meter). The Q of a resonant circuit characterizes the rate of change of the impedance near the resonant frequency. A coil inductor has both inductive and resistive components due to finite conductivity of the conductor. At resonance the spin system absorbs some of the RF energy, which affects the Q of the RF resonator, as the sample adds resistance due to its magnetic susceptibility. Let L_0 be the inductance of the empty resonator. If the sample susceptibility is χ, the loaded inductance becomes $L = L_0(1 + \chi)$. Including the resonator's self-resistance we have the following impedance:

$$Z = -i\omega L + r = -i\omega L_0(1 + \chi) + r \qquad (2.1.26)$$

Arranging it using the real and imaginary parts of χ:

$$Z = \omega L_0\chi'' + r - i\omega L_0(1 + \chi') \qquad (2.1.27)$$

Here $\Delta r = \omega L_0 \chi''$ is the sample resistance contribution to the resonator resistance. Q is defined as the ratio of the reactive impedance to the resistive impedance:

$$Q \stackrel{\text{def}}{=} \frac{Im\{Z\}}{Re\{Z\}} = \omega L/r \qquad (2.1.28)$$

If Q_0 is the Q of the empty resonator, inverse Q is increased by χ'':

$$\frac{1}{Q} = \frac{1}{Q_0} + \chi'' \qquad (2.1.29)$$

Since the change in $1/Q$ is very small, we can approximate the change as

$$\Delta Q = -Q^2\chi \qquad (2.1.30)$$

For the loaded detection circuit, in the absence of magnetic resonance, there will be a voltage $U = IR$ on the resonator where I is the continuously applied current. If B_0 is swept through the Larmor frequency of the sample, R changes from $Q\omega L$ to $(Q+\Delta Q)\omega L$, thus the resonator voltage changes from U to $U + \Delta U$ where

$$\Delta U = U\frac{\Delta Q}{Q} = -UQ\chi'' \tag{2.1.31}$$

Eq. 2.1.31 is only valid when the spin response linearly to the applied B_1. In general, although spins have a nonlinear response and complicated behavior, they can be assumed to be linear under certain conditions such as small flip angle excitation [41].

2.1.5.2.1 Cavity Resonator

In 1945, Purcell and coworkers used a resonant cavity in the form of a short coaxial line loaded heavily by the capacity of an end plate. The inductive part of the cavity was filled with paraffin resulting in a loaded Q of 670. The resonator was placed in the gap of the magnet, and RF power was coupled into the cavity at a level of about 10 W. The RF magnetic field in the cavity was everywhere perpendicular to the steady field. The residual signal, after amplification and detection, was measured with a micro-amperemeter. When the static magnetic field was slowly varied, a resonant absorption was observed which reduced the cavity output at resonance.

2.1.5.2.2 Geometrical Decoupling

In Bloch's experiment [32] two orthogonally placed solenoid resonators were used: one to apply the continuous oscillating RF field for RF excitation, the other to detect the NMR signal by Faraday induction. As a sample a small glass bulb filled with water was placed in the middle of the orthogonal resonator setup. This was the first detection setup that is based on Faraday's induction principle, rather than the Q-meter approach. The orthogonal placement is called geometric decoupling, and it provides a good isolation between the Tx and Rx resonators by reducing mutual inductance.

2.1.5.2.3 Rapid Scan Correlation Method: Frequency Sweep

Until Dadok and Gupta independently published the concept of rapid scan methods [8], [9], CW NMR with slow passage field sweep was already considered to be an inefficient way of spectroscopy. Dadok and Gupta suggested sweeping the RF pulse frequency rather than B_0. Choosing the sweep rate in the fast regime of the adiabatic passage they achieved a spin frequency response that is the cross correlation of a continuous wave spectrum with a single reference line recorded under the same conditions [8], [9]. This method replaced the conventional CW spectrometer by incorporation of a cross-correlator.

The swept RF pulse $b(t)$ can be described is a linear chirp pulse

$$b(t) = Ae^{-ict^2/2} \tag{2.1.32}$$

and the frequency response (Fourier transform) of the pulse is

$$B(\omega) = \int_{-\infty}^{\infty} b(t)e^{-i\omega t}dt = A\sqrt{\frac{2\pi}{c}}e^{-i(\pi/4-\omega^2/2c)} \tag{2.1.33}$$

In order to accumulate the fast linear sweep excitation spectra, a phase lock between the swept frequency and a constant reference frequency supplied to the synchronous detector is needed. The desired spectrum $H(\omega)$ is then obtained from the Fourier transform of the measured signal $S(\omega)$:

$$H(\omega) = \frac{S(\omega)}{B(\omega)} \tag{2.1.34}$$

In MRI, adiabatic pulses are typically used as a special class of frequency modulated (FM) RF pulses which have the ability to tolerate inhomogeneity in B_1 amplitude and resonance offset. Adiabatic pulses were introduced first to achieve uniform $\pi/2$ (adiabatic half passage) and π (adiabatic full passage) excitations with surface resonators [44]. In principle, rapid passage can be accomplished with any FM pulse that has an offset-independent adiabaticity [45]: chirp [46], hyperbolic secant (HS) pulse [47], and HSn pulses [45]. Such rapid passage pulses can be used to excite spins when satisfying the condition

$$\frac{d\omega_{RF}}{dt} \gg {T_2}^{-2} \tag{2.1.35}$$

where ω_{RF} is the time-dependent pulse frequency [9]. During an adiabatic passage, the magnetization is follows B_{eff} field as the amount of off-resonance changes and B_{eff} moves further away from the B_0 axis. The rate of change of frequency for adiabaticity condition is given by

$$\frac{d\omega_{RF}}{dt} \ll (\omega_1(t))^2 \tag{2.1.36}$$

where $\omega_1 = \gamma B_1(t)$. Rapid passage pulses on the other hand fulfill the condition

$$\frac{d\omega_{RF}}{dt} \gg (\omega_1(t))^2 \tag{2.1.37}$$

The conditions expressed in Eq. 2.1.36 and Eq. 2.1.37 are the adiabatic region and the linear region, respectively. The advantages of frequency modulated pulses operating in the linear region are [48]:

1. Linear region pulses excite a broad bandwidth using only low peak RF amplitude, because the bandwidth is set by the sweep range, not the peak power.
2. The bandwidth is determined by the range of the frequency sweep, and the excitation profile is highly uniform within the sweep range with certain types of rapid passage such as the HS and HSn. As a result, Fourier transform of the input RF signal is approximately constant in the baseband [49].

CEA methods use FM pulses in linear region and benefit from these advantages.

2.1.5.2.4 Recent Methods & Setups

In 2001, Davies and co-workers demonstrated the use of a hybrid coupler-based detection circuit in 2D CW rapid scan MRI of solids [11] (cf. section 2.1.6 for a brief description of image formation in MRI). The hybrid coupler was matched to 50 Ω, and one port was terminated with 50 Ω resistance while the other was connected to the matched birdcage resonator loaded with the sample of sub-millisecond T_2* value (Fig. 7). When the field was swept through resonance, due to NMR absorption a mismatch occurred at the resonator port. As a result of the reflection at input and output ports, they were able to detect the rapid scan correlation response to a chirp RF pulse at 7 T field strength. Lock-in amplification was used to increase the sensitivity of the detection.

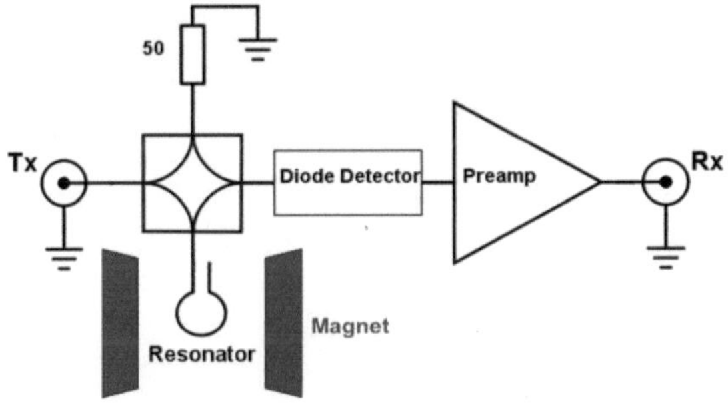

Fig. 7: Block diagram of a basic continuous-wave magnetic resonance spectrometer.

In 2003, this system was applied to imaging of heterogeneous materials [12], and later in 2005 upgraded to 3D CW MRI with rapid scan [13]. As the system consisted mostly of custom-made components, it took 10 hours to obtain a 2D image with sufficient quality.

Rapid scan or CW NMR techniques are adopted also in electron paramagnetic resonance (EPR) where the coherence times are much shorter than NMR [50].

Since the MRI research and practice have favored interleaved acquisition methods, MR hardware units also developed in that direction since end of 1970s. Until a few years ago, CEA using conventional MRI systems has been considered impossible. The fundamental challenge of implementing CEA is the 100 dB difference between the Tx RF pulse power and the received MR signal. To overcome this challenge, more recently several MR methods have been proposed:

Sweep Imaging with Fourier Transform (SWIFT)

The SWIFT technique uses FM RF excitation with **virtually** simultaneous signal acquisition [51]: in SWIFT, excitation is performed by a multiply gapped frequency swept RF pulse, e.g. a HS pulse, in the presence of readout gradient fields. Signal is sampled during each gap of the pulse, when the RF is off. The signal sampled throughout the pulse can be reconstructed, essentially by Fourier transform, into a profile along the direction of the gradient. SWIFT is limited by the Tx/Rx switching speed of the available RF circuitry, which, in practice, prevent approaching CEA operation and reaching high BW. In 2012, true CEA methods were demonstrated [52]–[54].

Sideband Excitation

Sideband excitation uses off-resonant excitation on the order of a few Megahertz, and filtering in the time domain using sharp RF filters [54]. Due to the size limitations of the setup, imaging of only very small objects was demonstrated (HS RF pulses with an offset frequency of 2.25 MHz). However, off-resonant excitation increases the RF power that is required to achieve flip angles comparable with on-resonant excitation. In addition, the sample will experience an unwanted Bloch-Siegert shift during off-resonant excitation [55], and the hardware required can be complex in clinical applications.

Continuous SWIFT

In 2012, Idiyatullin D, et al. demonstrated that it is possible to implement signal acquisition during RF excitation using a hybrid coupler-based decoupling scheme between Tx and Rx resonators [53]. In this work, isolation was high enough to reduce Tx induced signals in the Rx resonator to the Rx dynamic range. The hybrid coupler creates a 180°phase difference between the RF input port and the output port to the receiver circuitry. Thus, the hybrid coupler subtracts the unwanted input RF excitation signal from the acquired signal, which is the additive combination of the MR signal and the RF excitation, so that only the desired MR signal is received. Due to nonidealities in the circuit and the frequency dependency of the hybrid coupler, the unwanted RF excitation signal often leaks through to the receiver due to the limited isolation (maximum of 30-40 dB isolation was reported in [53]). Thus, the true MR signal needs to be extracted from the acquired data by a cross-correlation technique.

The hybrid coupler based decoupling has several additional disadvantages:

1. The amount of the decoupling is low.
2. High dynamic range receiver electronics is required.
3. Extremely accurate tuning of the isolator is needed.
4. Decoupling is sensitive even to small changes in resonator impedance.
5. The acquired signal is more sensitive to the imperfections of the RF excitation field.

2.1.6 Image Reconstruction

In MRI, linearly varying magnetic fields, so called gradients are employed for spatial encoding. A gradient field, G(t), along x direction combined with the B_0, will result in position dependence of the Larmor frequency:

$$B(x,t) = B_0 + xG(t) \tag{2.1.38}$$

$$\omega(x,t) = \omega_0 + \omega_G(x,t) \tag{2.1.39}$$

for

$$\omega_G = \gamma x G(t) \tag{2.1.40}$$

If MR signal is acquired in the presence of a gradient, the spatial information is encoded in the frequencies (frequency encoding). Frequency encoding in two dimensions was applied in 1971 by Lauterbur, who used CEA methods to acquire the first MR image [6]. Lauterbur used CW NMR acquisition in his experiment. A few years later Ernst showed that similar encoding strategy is applicable also to interleaved acquisition methods [14]. Lauterbur named this imaging method *'zeugmatography'*. With a pure frequency encoding method projections of the object are acquired under different angles, and a similar reconstruction method such as filtered back projection (Fig. 8) can be applied as in computed tomography (CT).

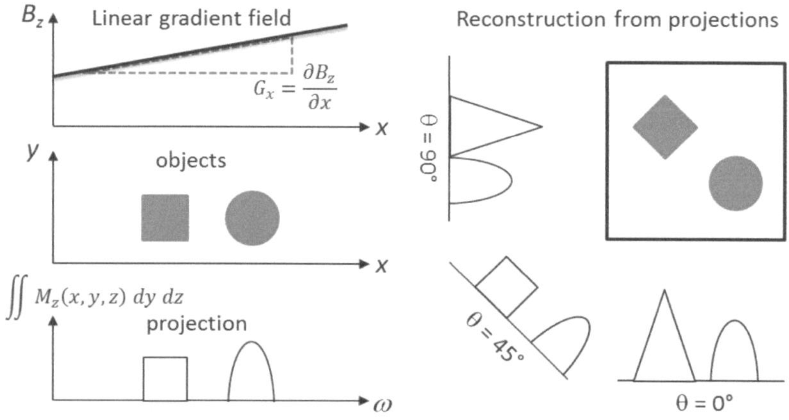

Fig. 8: Demonstration of frequency encoding applied to projection reconstruction in MRI.

In 2D case, if $g_\theta(x')$ is the projection of $f(x,y)$ at the angle θ, and $F(k_x,k_y)$ is the Fourier transform of $f(x,y)$; the central slice theorem states that the 1D Fourier transform of the projection $g_\theta(x')$ is equal to 2D Fourier transform $F(k_x,k_y)$ along a radial line at the angle θ [33]. For $\theta = 0$, this can be seen from:

$$F_{1D} = \int_x \left(\int_y f(x,y)dy \right) e^{-i2\pi k_x x} dx =$$

$$= \int_x \int_y f(x,y)\, e^{-i2\pi(k_x x + 0y)} dxdy \;= F(k_x, 0)$$

(2.1.41)

As a result, each FID gives a line at the projection angle θ in the 2D Fourier transform of the object, which can be used to reconstruct the image.

This reconstruction concept can be extend to 3D frequency encoding, which will later be used in ultra-short echo time (UTE) and the CEA sequences. In MRI, Fourier space is also called k-space. In 3D radial MRI, the x-, y-, and z-gradient fields for frequency encoding are applied in different combinations to cover the whole 3D k-space. Saff formulated equally distant points on a surface of the sphere resulting in a set of angles θ_n and $\emptyset_n$ (spokes) [56]. Starting from the center of the k-space (*k* = 0), data are acquired along radial spokes up to a maximum value k_{max}, and the direction of the spokes is varied from one acquisition to the next to cover the surface of a sphere (n=1:N_{spoke}):

$$k_x(n) = G_{max} \sin(\theta_n) \cos(\emptyset_n)\, k \tag{2.1.42a}$$

$$k_y(n) = G_{max} \sin(\theta_n) \sin(\emptyset_n)\, k \tag{2.1.42b}$$

$$k_z(n) = G_{max} \cos(\theta_n) k \tag{2.1.42c}$$

where,

$$h_n = -1 + \frac{2(k-1)}{(N_{spoke}-1)} \quad , \quad 1 \leq n \leq N_{spoke} \tag{2.1.43a}$$

$$\theta_n = \arccos(h_n) \qquad (2.1.43b$$

$$\emptyset_n = \left(\emptyset_{n-1} + \frac{3.6}{\sqrt{N_{spoke}}} \frac{1}{\sqrt{1-h_n^2}} \right) (mod\ 2\pi) \tag{2.2.43c}$$

$$\left(2 \leq n \leq N_{spoke} - 1, \emptyset_1 = \emptyset_{N_{spoke}} = 0\right)$$

In projection imaging, the radial k-space samples are non-uniform: the k-space center is denser, and the periphery is sampled at (or even below) the Nyquist frequency. To reconstruct the 3D image data by the Fourier transform, the radially sampled data must be resampled to a Cartesian grid.

This problem is resolved by 're-gridding' method, which was introduced in CT by O'Sullivan [57], and adapted to MRI by Noll in 1991 [58]. Re-gridding estimates the sample values at the Cartesian grid positions based on the information obtained from non-uniform samples. Each non-Cartesian sample has a certain contribution to the signal supposed to exist at the Cartesian grid. In this thesis, Kaiser-Bessel interpolation was used to calculate the contribution of non-Cartesian samples on each grid. Finally a density correction is applied to account for the dense sampling at the center, as well as apodization correction is employed to form the re-gridded k-space (Fig. 9).

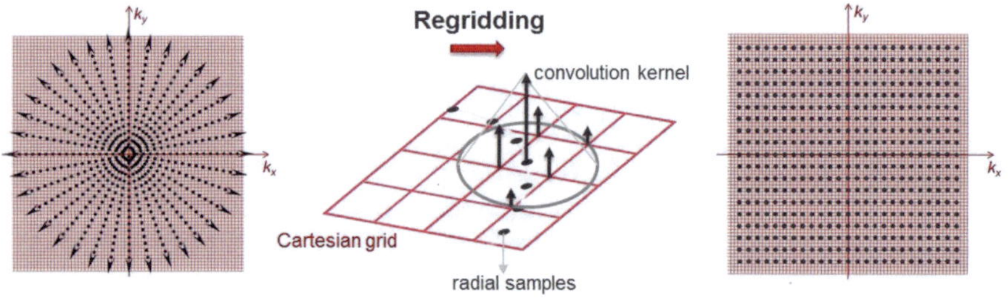

Fig. 9: Schematic of the re-gridding operation. The data is acquired as sequential radial spokes: dense in the center, sparse in periphery. Data values corresponding to each Cartesian grid points is calculated by interpolation of radial sampling points on to the Cartesian grid based on the selected convolution kernel. Finally, Cartesian k-space with uniform sampling is formed with proper density correction

2.2 MRI Hardware

A standard whole body MRI system consists of a superconducting magnet to provide the static field B_0, linear gradient coils (G_x, G_y, G_z) for spatial encoding, and a whole body RF resonator to provide the RF field $B_1(t)$ for homogenous excitation of the spins. Local Rx RF resonators are used to acquire MR signal only from a specific region of body. RF resonators and the related electronic circuitry will be explained in detail in this section.

2.2.1 RF Resonators: Antennas for MRI

In principle, RF resonators (also called '*RF coils*') in MRI are near-field narrow-band resonant antennas. Tx RF resonators are used to deliver the excitation magnetic field, B_1 to excite the spins, and the Rx RF resonators to receive the induced MR signal and to transfer it to the Rx circuitry. Purcell's re-entrant cavity resonator (section 2.1.5.2.1.) and Bloch's solenoid resonator pair (section 2.1.5.2.2.) were the first examples of Tx/Rx RF resonators for MR, and since then, RF resonators evolved to more sophisticated coil designs and materials to ensure that an optimal SNR is achieved in MRI experiments.

In the next section the concept of impedance matching for RF resonators will be introduced, making use of the quality factor, *Q*, as defined in section 2.1.4.1. Here, *Q* will be formulated and its relation to the signal to noise ratio (SNR) of the measurement will be discussed.

2.2.1.1 Impedance Matching, Q, and SNR

A general schematic model of the analog part of an MR signal detector, i.e. Rx resonator, is given in Fig. 10. It consists of the resonator part, the RF resonator, which is tuned to the Larmor frequency, and a partially balanced matching unit to match the impedance to 50 Ω. For the Rx resonator arrays that are widely used in clinical MRI units, impedance is not necessarily matched to 50 Ω, but to the input impedance of the subsequent preamplifier. A 50Ω-coaxial cable then carries the signal to the preamplifier. At the preamplifier, the MR signal is amplified by 20-35 dB and then sent to the analog to digital converter (ADC) via baseband detection unit, where it is digitized for further processing.

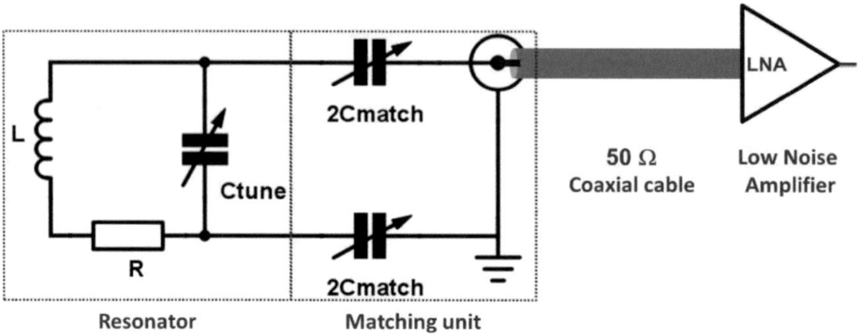

Fig. 10: A lumped element model of the RF resonator and the matching unit for an MR signal detector unit. Instead of a single matching capacitor of capacitance Cmatch at the inner conductor, capacitors of 2Cmatch are connected to both conducting lines for partially balanced matching.

The purpose of impedance matching is to bring the resonator impedance to a purely resistive 50 Ω so that the reactive inductance of the resonator is compensated and the remaining series resistance of the resonator is converted to $Z_0 = 50\ \Omega$. This is achieved by using a tuning and a matching capacitor. For matching, one capacitor C_{match} at the inner conductor would be sufficient, but this would cause unbalanced currents and standing waves on the shield of the coaxial cable leading to inefficient power transmission, coupling of the unbalanced fields to other conducting parts, and eventually heating of the cable and the surrounding conductive elements. To avoid unbalanced currents, balanced matching with 2 capacitors was suggested [59]. To further reduce unbalanced currents, cable traps or balanced-unbalanced transformers (BALUNs) are used. For details of construction of these, please refer to Appendix A.

2.2.1.1.1 Q of the Resonator

The quality factor, Q characterizes the rate of change of the impedance around the resonant frequency. Q is 2π times the ratio of the RF energy stored in the RF resonator and the energy lost over one period of oscillations:

$$Q = 2\pi f \frac{W}{P} = \omega \frac{W}{P} = \omega \frac{LI^2/_2}{RI^2/_2} = \omega \frac{L}{R} \tag{2.2.1a}$$

$$Q = \frac{\omega}{\Delta\omega} = \frac{f}{\Delta f} \tag{2.2.1b}$$

Here, P is the RF power delivered to the RF resonator. The Q factor can be measured using a network analyzer's reflection measurement: the resonant frequency f divided by Δf (3 dB bandwidth) yields Q. Half the RF power is consumed at the frequencies corresponding to 3 dB less than the resonant peak power in a reflection measurement. Δf is the difference between these two frequencies.

2.2.1.1.2 Q and Ring-down Recovery Time

The ring-down time of a resonator describes the damping of the oscillating energy in the resonant circuit after blocking the applied RF power. Natural ringing time constant is given by $2Q/\omega_0$. Following this definition, the recovery time of a probe, T_{Rec} is:

$$T_{Rec} = \frac{2Q}{\omega} \ln \frac{U_{B_1}}{U_{MR}} \tag{2.2.2}$$

where, $U_{\mathrm{B_1}}$ is the peak RF pulse amplitude, and U_{MR} is the MR signal induced peak voltage amplitude [60], [61]. For a RF resonator with Q = 100 at 64 MHz, if we assume a 100 dB difference between the RF pulse and the MR signal levels, T_{Rec} = 5.7 μs.

2.2.1.1.3 Q and Signal to Noise Ratio (SNR)

Q is related also to the sensitivity and the loading condition of the resonator. If we assume a homogeneous RF field and a uniform sample with volume V_{S}, the Q of the loaded resonator amounts to

$$Q = \omega \frac{W}{P} = \omega \frac{\int_V B_1^2 dV}{P} \frac{\int_S B_1^2 dV}{\int_S B_1^2 dV} = \omega \frac{\int_S B_1^2 dV}{P\eta} = \omega \frac{B_1^2 V_S}{P\eta} \tag{2.2.3}$$

The efficiency of the resonator is expressed by the ratio of the effective B_1 over the input B_1, which is proportional to $\sqrt{P}$:

$$\frac{B_1}{\sqrt{P}} = \sqrt{\frac{\eta Q}{\omega V_S}} \propto \sqrt{\eta Q} \tag{2.2.4}$$

Here, η is called the *filling factor* of a resonator by the sample volume, and describes the RF energy delivered to the sample over the entire energy of the resonator. As Q and η increase, efficiency of the resonator increases.

Noise

The noise source in MRI is called thermal noise or Johnson noise and originates from the random motion and interaction of charge carriers in a conductor or Brownian motion of electrolytes in a sample. As the e.m.f. induced by the spins create a detectable NMR signal, thermal noise e.m.f. also causes a noise signal which is detected together with the NMR signal. Nyquist formulated thermal noise power concept in 1928 [62]: the time domain root mean square noise e.m.f. is

$$\sigma = \sqrt{4kT\Delta fR} \tag{2.2.5}$$

where T is the absolute temperature in °K, Δf is the frequency bandwidth which depends on the band-pass filter at the receiver electronics, and should be set to cover the sampling bandwidth. R is the equivalent series resistance of the loaded resonator. R represents the resistance in both the RF electronics, including the resonator and the body. In whole body MRI at clinical field strengths the contribution from electronics and the resonator to R is much smaller than that of sample (e.g., the patient) especially when the conductivity of the load is high. Patient noise e.m.f. is a result of random RF currents circulating round a number of RF eddy current loops. The noise from individual sources, such as the resonator or the patient, can be assessed by calculating the power deposition occurring when the resonator is used to transmit [63].

During signal reception, small local Rx resonators are advantageous to reduce noise e.m.f. since the amount of loading within the sensitivity region is less than with a large resonator covering the whole body. Another controllable parameter is the receiver bandwidth: in an MRI acquisition Δf is inversely proportional to the total acquisition time t_{ADC} (the time during which the ADC is switched on) and the MR signal is acquired. In case of rapid T_2* decay, reducing Δf is not preferable to improve the noise, because the acquired signal decays rapidly. As a result, most of the ADC sampling points do not contain information but noise. Besides, the sampled data is strongly weighted due to the exponential decay.

SNR

Following Hoult's [31], [40], [64] signal formulation based on the principle of reciprocity and using Eq. 2.2.4, we can directly relate η and Q to SNR:

$$|S| = \frac{\omega\omega_0 B_1 M_0}{I} = \frac{\chi\omega_0^2 B_1 V_S}{\gamma\mu_0 I} \tag{2.2.6}$$

$$SNR = \frac{\chi}{\gamma\mu_0\sqrt{4k_B T\Delta f}}\frac{\omega_0^2 B_1 V_S}{I\sqrt{R}} \sim \frac{\omega_0^2 B_1 V_S}{I\sqrt{R}} = \omega_0^2 V_S \frac{B_1}{\sqrt{P}} \tag{2.2.7}$$

Using Eq. 2.2.4 we get

$$SNR = \omega_0^{3/2} V_S^{1/2}\sqrt{\eta Q} = \sqrt{\omega_0^3 V_S \eta Q} \tag{2.2.8}$$

Eq. 2.2.8 suggests that for a single RF resonator, optimizing SNR is the same as optimizing the Tx efficiency (cf. Eq. 2.2.4). Changes of Q due to loading need to be considered to evaluate the noise contribution of the resonators. As a figure of merit for evaluation of noise figure, NF of a resonator the noise power ratio, NPR is defined [65]:

$$NPR = \frac{R_S + R_C}{R_S} = \frac{Q_U/Q_L}{Q_U/Q_L - 1} = \frac{SNR_{power}(unloaded)}{SNR_{power}(loaded)} \tag{2.2.9}$$

$$NF_{coil} = 10\log(NPR)[dB] \tag{2.2.10}$$

Here, the resonator noise figure, NF measures the SNR reduction attributed to the losses in the resonator. R_S and R_C are the sample and resonator resistances, respectively. The reduction in voltage SNR is proportional to $\sqrt{\mathrm{NPR}}$. NF of 1 dB corresponds to 10% reduction in SNR. When $R_S >> R_C$, the sample noise is dominant. To minimize R_C, inductive elements and capacitors with low effective series resistances (ESR) are preferred.

RF resonators can be used as Tx-only resonators, as Rx-only resonators, or as Tx/Rx resonators depending on the electrical design. In the following subsections, type of the RF resonators that are used in this thesis are described.

2.2.1.2 Surface Loop Resonators

Surface resonators were first used and are still used in oil wells before they were introduced to in vivo MR field [66]. Although they were introduced in 1980 as "unusual type of RF coils" by Ackermann [67], surface resonators have become the most commonly used resonators in MRI. Typical loop resonator composes a wire or copper tape to form the inductive part where magnetic field is generated, capacitors for tuning to the Larmor frequency, and impedance matching to the transmitter or noise matching to the receiver. Penetration depth of the magnetic field generated by a surface loop resonator is approximately equal to the radius of the loop [67].

A practical hint for designing most RF resonators for MRI is to distribute the capacitance over the inductance as shown in Fig. 11. A rule of thumb is to keep the length of the inductive wire segments between $\lambda/10$ and $\lambda/20$ to make sure that RF fields do not deviate along the wire, where λ is the wavelength. A wire segment of λ causes the RF field to cancel each other at certain positions (Fig. 11, left). Distributing capacitance helps also reducing the electric field formed around a capacitor, and thus the energy stored on it.

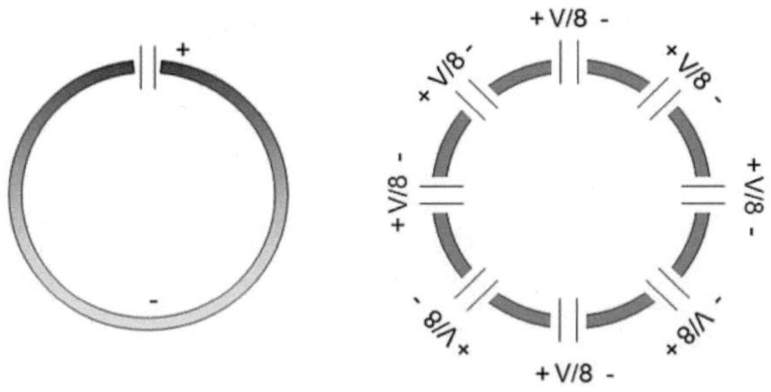

Fig. 11: Demonstration of the effect of distributing capacitance over the loop resonators. In single capacitor design (left) the magnetic field cancels out in the middle and the power on the capacitor is 64 times the power on each capacitor in the distributed capacitor design (right).

2.2.1.3 Solenoid Resonators

Solenoid resonators are basically cylindrically shaped air core inductors and they generate longitudinal magnetic field along their core axis. They are often considered to have the best performance and widely used in MR spectroscopy and preclinical MRI systems due to SNR advantage. In [40], Hoult states that the sensitivity to MR signal is proportional to the intensity of the magnetic field at the position of the signal source. Solenoid, from this perspective, is the optimal case where all the current carrying wires are close to the load.

A disadvantage of the solenoid resonator is the B_1 direction, which may lead to B_0 homogeneity and sample accessibility problems when placed perpendicular to the B_0 direction in whole body superconducting magnets. Solenoid design is not preferable for large samples as the resonator inductance increases rapidly with its dimensions. B_1 field homogeneity is satisfactory when the length of the resonator is sufficiently long, which however contradicts with the tendency to keep inductance low to avoid self-resonance at unwanted frequencies.

Magnetic field efficiency for solenoid is given in [68] as

$$\frac{B_1}{\sqrt{P}} = F_{geo}\sqrt{\frac{Q}{\omega}} \quad (2.2.11)$$

where, F_{geo} is the geometry factor that depends on the dimensions of the resonator. As long as the self-resonance frequency is well above the tuning frequency, Q depends only on the volume of the resonator and the square root of the tuning frequency. As the diameter and the length increase, the resonator impedance will also increase, which means that the number of turns should be decreased. The limiting case of single-turn solenoid corresponds to a loop-gap resonator design.

2.2.1.4 Loop-gap Resonator

Loop-gap resonator [69] is a special type of solenoid resonator. Many variants of the loop-gap resonators exist and they are used frequently in high-frequency applications. In a loop-gap resonator, the current density along the conductor is almost constant and increases rapidly at the edges. Although the magnetic field decreases towards edges, the increase in current density compensates for the reduction in the magnetic field. It is thus very homogeneous in the longitudinal axis, superior to the solenoid. Q is generally comparable with that of solenoid resonator of similar dimensions. The loop-gap resonator is also more efficient especially in loaded case due to the reduction of dielectric losses, as it has lower electric fields. This property makes the loop-gap resonator useful for the electric field-critical applications.

A disadvantage of the loop-gap resonator in large scale applications are eddy currents that are formed on the wide area conductor surface due to changing magnetic field gradients. This can be substantially reduced by galvanically separating the conductor into smaller parts using DC blocking capacitors. It is also possible to distribute the tuning capacitors serially.

2.2.1.5 Birdcage Resonators

A birdcage resonator consists of conducting legs (or rungs) and end rings which are interleaved by tuning capacitors (Fig. 12). Birdcage resonators are designed to produce linearly or circularly polarized, highly uniform transverse RF magnetic fields within a cylindrical volume. It was found in 1883 by analyzing the fields of a uniformly magnetized cylinder that a current density distributed sinusoidally on the surface of a cylinder yields a uniform transverse magnetic field [70]. A transverse magnetic field in a volume resonator is advantageous, as the cylindrical openings align with the B_0 while B_1 field is vertical to B_0 (Fig. 12a). Moreover, birdcage resonator self-resonance frequency is well above the typical application frequency range, therefore increasing resonator dimensions will not degrade the resonator performance significantly. Birdcage resonators are preferred in clinical whole-body MRI systems over solenoid or loop-gap resonators. Birdcage resonators can be designed to have a low pass, high pass or band pass frequency response. At clinical field strengths for ^{1}H, low pass designs are preferable as they can be tuned with commercially available capacitors. The equivalent lumped element network of a hybrid design birdcage resonator is shown in Fig. 12b.

I_n are the currents flowing in the legs of the birdcage. To create a homogenous magnetic field, we need the current on leg n need to fulfill the following equation (written in phasor notation):

$$I_n = \frac{I_0 e^{i2\pi n}}{N} \tag{2.2.12}$$

where, I_0 is the maximum current, and N is the number of legs.

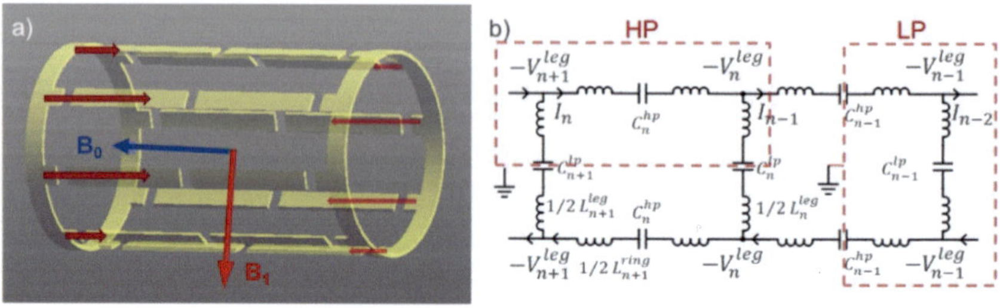

Fig. 12: Birdcage resonator and lumped element representation. (a) Schematic of birdcage leg and end-ring connections and representation of the leg current distribution resulting in a transverse magnetic field vertical to B_0. (b) Lumped element equivalent of a birdcage resonator for a hybrid design with capacitive elements in legs and end rings.

For a given geometry and materials with known properties it is possible to estimate the inductance and calculate an approximate value for each capacitance by solving the lumped element circuit equation [71]. Birdcage resonators have $N/2$ modes resulting from various phase-combinations of the current elements. Each mode has a unique resonance frequency. The homogeneous mode is the only mode with a positive magnetic field at the center of the cylinder. The other modes have zero magnetic field at the center but high field near the legs. For a low-pass birdcage resonator with capacitive elements soldered at the rungs, the homogeneous mode has the lowest frequency, whereas it has the highest frequency for a high-pass design. In practice, the birdcage resonance can be tested with a wideband pick-up loop and S11 measurement using a network analyzer. In S11 measurements, the homogeneous mode experiences the least change when the pick-up loop is moved from edges towards the center.

If the birdcage is driven at one port it generates a linearly polarized magnetic field. It is also possible to excite and detect in quadrature mode, which increases the SNR by $\sqrt{2}$ [72]. To generate a circularly polarized magnetic field, two ports separated by 90° are driven with 90° phase difference. A 90° hybrid coupler is used to drive both ports simultaneously from a single RF source (cf. section 2.4.2.1.1).

Symmetry is crucial in birdcage design to ensure the desired sinusoidal current distribution. Practically, this can be achieved by a precise mechanical design and using capacitors with tolerance range of at most 1%. Taking the non-ideal nature of the hand work into account in building such resonators, some degree of freedom for fine-adjustments of isolation, matching and tuning is provided using variable capacitors. In quadrature design, variable tuning capacitors are placed at the symmetrical legs of the driving ports to ensure that tunings of both ports do not interact [73]. For isolation, a variable capacitor is connected at equal distance to both ports. At least 20 dB isolation is required for proper function of birdcage. For the matching, a variable capacitor is connected in parallel to the matching capacitors at the driving ports. To match one port, the other port is left open by dismounting the leg (low-pass) or end-ring (high-pass) capacitor. In this way, both ports are brought to a similar matching condition before fine tuning. An example placement of ports and variable capacitors is shown for a high-pass birdcage resonator in Fig. 13.

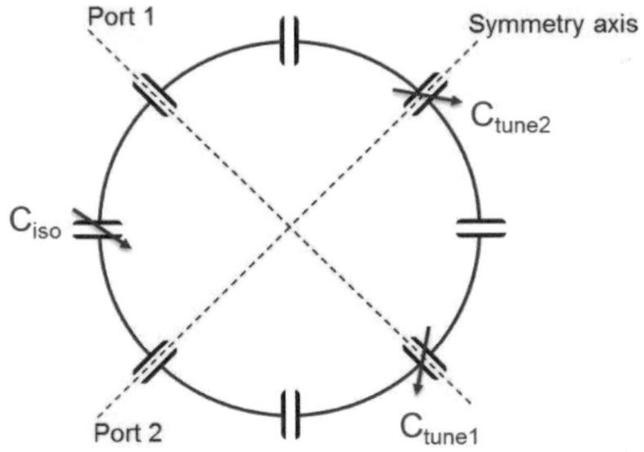

Fig. 13: Demonstration of tuning and isolation capacitor positions on an 8-leg high-pass quadrature birdcage resonator shown in axial plane.

2.2.1.6 B_1 Field Characterization of RF Resonators

The excitation flip angle in MRI varies across the imaging volume due to inhomogeneity of B_1. Precise measurement of flip angle is important for correction of spatial variations in image intensity, quantitative data analysis, and parallel transmission in MRI. Several methods have been proposed to extract spatial distribution of flip angle:

1. Gradient recalled echo (GRE) double-angle method: two acquisitions are performed with a spoiled GRE sequence (FLASH) with different nominal flip angles [74]–[76]
2. Actual flip angle imaging (AFI): two acquisitions interleaved in a GRE sequence with equal flip angle but different TRs [77].
3. Phase-sensitive method: the phase difference between two acquisitions is used a measure of actual flip angle [78].
4. Bloch-Siegert (BS) based B_1 mapping technique [79]: An off-resonance RF pulse adds phase to the excited spins. For a large off-resonance frequency, the added phase is directly proportional to the square of $|B_1|$ [80].

Due to the applicability in samples with short T_2, the double-angle method was used in this thesis. If the TR is sufficiently long, the transverse magnetization (eqn.2.1.25), hence the signal intensity will be proportional to $\sin\alpha$ in a GRE sequence. If the same measurement is repeated with a flip angle of 2α, α can be calculated from the resulting image data, simply dividing the magnitude of the images:

$$\frac{M_{xy2}}{M_{xy1}} = \frac{M_{SS}\sin 2\alpha}{M_{SS}\sin\alpha} = 2\cos\alpha \tag{2.2.13}$$

$$\alpha = \cos^{-1}\left(\frac{M_{xy2}}{2M_{xy1}}\right) \tag{2.2.14}$$

B_1^+ at the required position is then calculated from the shape of the RF pulse using [81]

$$\alpha(\vec{r}) = \gamma \int_{-T_{RF}/2}^{T_{RF}/2} B_1^{+}(\vec{r},t)dt \qquad (2.2.15)$$

These equations are only valid for small flip angles, i.e. $\alpha < \pi/3$.

2.2.2 Transmit and Receive Chain

General outline and main elements of the Tx and Rx circuitry in MRI is shown in Fig. 14.

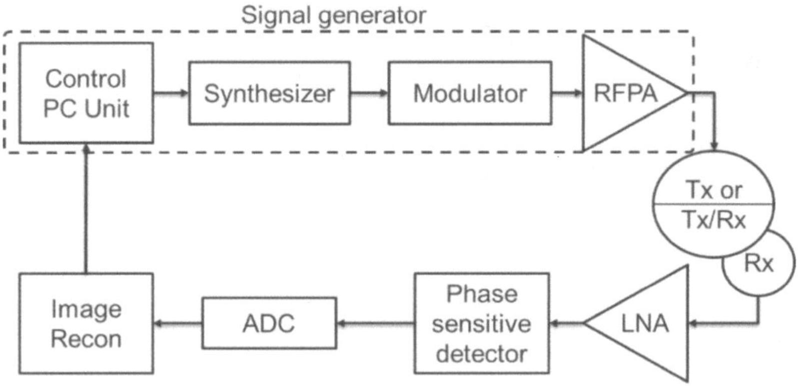

Fig. 14: Main elements of the Tx and Rx circuitry in MRI systems.

2.2.2.1 Signal Generator

Signal generator is composed of an RF synthesizer, a modulator and an RF power amplifier (RFPA), which are controlled by the system's computer unit. Here, the shape, duration, and phase of the RF pulse are determined. Synthesizer employs precise local oscillators and generates the sinusoidal RF signal with the carrier frequency equal to the Larmor frequency. The modulator applies the required amplitude and phase modulations to the RF signal. Then, the signal with adjustable peak amplitude (typically from mV to few Volts range) is sent to the RFPA with fixed gain of typically around 70 dB.

Although RFPAs are indispensable components of the interleaved MRI applications where there is need for RF pulses of high peak power, there is little research conducted in improving the RFPAs. This is probably due to the fact that the excitation and acquisition events are separated in time, and many specifications related to Tx noise, and stability are relaxed in interleaved acquisition methods.

Ideally RFPAs generates the exact replica of the input signal with the peak amplitude amplified. However, the most commonly used class-A RFPAs distort the signal invariably. Some of the important specifications for RFPAs are the following:

- blanked noise voltage
- unblanked noise voltage
- bias enable voltage spikes
- rising transition duration
- overshoot ringing/decay time
- falling pulse overshoot
- amplitude and phase stability and linearity

Detailed descriptions of those and more can be found in literature [82]. Most relevant to this study is the unblanked and blanked noise voltages, and the voltage spikes. Since class-A RFPAs have high noise figures and create an extreme amount of noise power during their active state, they are blanked during Rx by reverse biasing the transistors. Therefore, they are unusable for the CEA applications they are unusable. In our experiments, RFPAs are replaced by custom made low power amplifiers.

2.2.2.2 Receiver Chain

2.2.2.2.1 Pre-amplification

Receiver circuitry starts with pre-amplification of the MR signal in the Rx resonators. Preamplifiers used in MRI are low noise amplifiers (LNA). LNAs are designed to have low noise contribution so as to keep SNR of the acquired signal high. The key factor in designing LNAs is to use a transistor with low noise factor (F), optimizing the source resistance according to the noise voltage model of the transistor, and to minimize F in particular in the first stage of the amplifier, as the total F of a multi-stage LNA, F_{total} is dominated by the F of the first stage. The total gain of the amplifier should be set to the minimum necessary value for a good noise performance [83], as an excessive gain would restrict the available dynamic range for signal detection. The minimum LNA gain G_{min} can be determined by the noise factor, F_R of the rest of the receiver using the Friis equation [84]:

$$F_{total} = F_{LNA} + \frac{F_R - 1}{{G_{min}}^2} \tag{2.2.16}$$

According to this equation, if F_R is 10 dB and only 0.05 dB degradation in SNR is allowed, (G_{min} + F_{LNA}) should be 28.9 dB. In conventional MRI systems, there is an attenuator at the receive part which can be set to prevent the receiver from being saturated.

2.2.2.2.2 From Analog to Digital Domain

Digitization of the amplified MR signal is done by a 16-bit ADC and followed by a chain of digital filters to enhance the quality of the digitized signal. Note, that to avoid aliasing in ADC sampling, analog filters are also applied to remove any frequency higher than the maximum allowed Nyquist frequency before the analog signal is digitized. A 16-bit digital encoding of the analog signal allows 2^{16} discrete steps to represent the analog voltage levels. If an amplitude level stays between two discrete voltage levels, a rounded version of it is represented in the digital data. This is called quantization error which leads to white noise and distortion of weaker signals resulting in poor SNR and low dynamic range [85]. The number of bits has a direct effect on the SNR:

$$SNR = (6\ \mathrm{dB} \cdot number\ of\ bits) - 1.1\ \mathrm{dB} \tag{2.2.17}$$

Oversampling with ADC at a higher frequency than Nyquist frequency also helps improving precision. A four-fold oversampling reduces the quantization noise by a factor of 4, as the noise is spread over a larger spectral range [86].

Various digital filters are applied to the successfully digitized data [85], details of which are beyond scope of this thesis. One implication of digital filters is the acquisition delay. Both analog and digital filters experience group and phase delays. For both type of filtering, the longer the group delay, sharper is the frequency response at the transition band. Although, in theory, digital filtering can be optimized for a specific application, conventional MRI systems do not allow modification of the filters at the receive chain. Therefore, the user is limited by a hard-coded acquisition delay. However, digital filtering can be improved using the state-of-the-art real-time filters together with advanced digital processor units to minimize the acquisition delay.

Quadrature Detection

Quadrature detection is used in conventional MRI Rx chain. The RF signals at frequency f coming from the resonator are mixed with (f_0 + f_0') to produce an intermediate frequency f_i using a double balanced mixer. Once digitized, the rotating frame signals $M_x(t)$ and $M_y(t)$ are created using a digital quadrature detector and a digital filter. This is basically a computation in software level. The quadrature demodulation can also be employed in the analog mixing stage in order to obtain the in-phase and quadrature components of the signal using two phase sensitive detectors. The first detector has a reference sinusoidal signal oscillating at the Larmor frequency generated by the synthesizer, whereas the second detector has a reference sinusoidal signal shifted by 90°. The first and the second phase sensitive detectors allow to obtain the 'real' and the 'imaginary' signals, respectively.

The intermediate frequency and digitization is typically at 1 MHz, resulting in oversampling and the generation of too much data to be conveniently stored. Digital filtering eliminates the high frequency components from the data, and decimation reduces the size of the data set.

2.2.2.2.3 Active Detuning of Rx Resonators

During RF excitation, Rx resonators should be decoupled from the Tx resonators to avoid damaging the Rx circuitry. This is accomplished by intentionally shifting the resonance frequency of the Rx resonators during Tx (active detuning). Active detuning reduces mutual coupling between Tx and Rx resonators, so that the Tx efficiency is not compromised. A standard active detuning circuit layout is shown in Fig. 15.

The most common active detuning method is to connect an arbitrary inductor in series and a PIN diode in parallel to one of the tuning capacitors in the resonator. PIN diodes have a large intrinsic semiconductor layer (*I*) between P and N-doped layers. When forward biased, charge stored in *I*-layer allows diode to conduct. When reverse biased, the stored charge needs to be removed which depends on the reverse recovery time of the diode. The PIN diode in Fig. 15 is activated via a DC bias voltage carried along the inner conductor of the coaxial cable, and the inductance of the resonator increases by addition of the extra inductance, thus, the resonance shifts to a higher frequency.

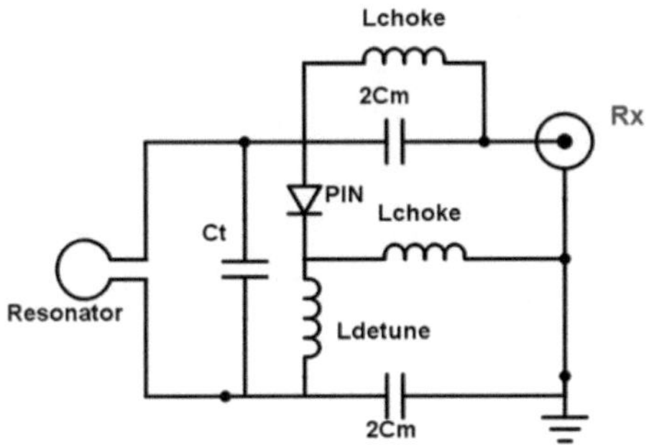

Fig. 15: PIN diode based active detuning circuit. When the diode is forward biased, the resonance shifts to a higher frequency due to increased inductance by Ldetune, thus the Rx resonator is decoupled from Tx resonators during RF excitation.

Passive detuning methods using fast switching crossed diodes or back-to-back Schottky diodes are also used in Rx only resonators. Diodes automatically break through when an alternating current greater than the turn-on voltage is applied. Passive detuning helps reducing number of elements, and simplifies the design eliminating the need for a DC voltage delivery. The expected resonance shift in detuned state should at least reach a few MHz.

2.2.2.3 Tx/Rx Switching

In MRI, Tx/Rx resonators are used when there is need for local resonators to increase Tx efficiency. Active detuning cannot be used as the resonator should be resonant both during Tx and Rx. Instead, Tx/Rx switches are employed to exchange the electrical path between resonator-RFPA, and the resonator-preamplifier electronically (Fig. 16). Tx/Rx switches protect the Rx circuitry, as well as isolate the Rx resonators from the noise generated by the blanked RFPA during Rx operation. Performance criteria for a standard Tx/Rx switch are

- high power handling capability
- low insertion loss
- high isolation
- high switching speed

Although, there are various type of Tx/Rx switch designs such as the classical crossed diodes switch [61] and transistor based switches [87], PIN diode switches are the most widely used Tx/Rx switches in MRI systems. Selection of PIN diode is essential in Tx/Rx switch design. Working frequency range, bandwidth, power handling capability, carrier life time, forward series resistance, thermal resistance are the main parameters for choosing a suitable PIN diode. Primarily, the switching speed is effected by PIN diode's transition time from forward to reverse bias, T_{FR}:

$$T_{FR} = \tau \ln\left(1 + \frac{I_F}{I_R}\right) \quad (2.2.18)$$

where, τ is carrier lifetime, I_F is forward bias current, I_R is initial reverse current. τ increases with increasing width of the *I*-layer, *W*. Forward series resistance, thus power dissipation and insertion loss, also increase as *W* becomes larger.

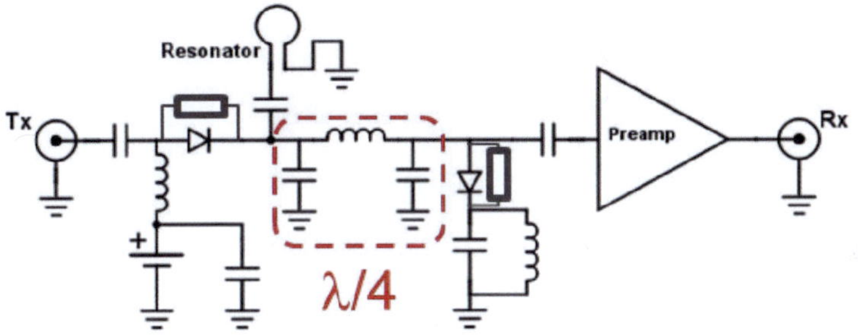

Fig. 16: Basic linear Tx/Rx switch. λ/4 line is replaced by lumped elements. During Tx mode, PIN diodes are forward biased, thus the RF signal from the Tx port is sent to the resonator. In Rx mode, diodes are reverse biased blocking the RF path between the Tx and Rx port, thus the received signal from the resonator is sent to the Rx port via a preamplifier.

Note that in reverse biased state PIN diodes behave like a resistor and a small parallel capacitor (total capacitance, C_T). In designing with PIN diodes, C_T should be taken into account especially if the PIN diode is serially connected to a resonant circuit. An example circuit for PIN diode based Tx/Rx switch is shown below.

In Tx mode, PIN diodes are forward biased. The RF path between the Tx port and the resonator is available, whereas the Rx path is blocked by combination of the $\lambda/4$ line and the shunt diode. $\lambda/4$ line acts as a high impedance against the Tx signal by transforming the low resistance created by the forward biased shunt diode to a high impedance. Thus, the RF signal from the Tx port is sent to the resonator. In Rx mode, diodes are reverse biased blocking the RF path between the Tx and Rx port. $\lambda/4$ line transforms the high impedance created by the reverse biased diode to a low impedance, allowing the RF signal to pass through. Thus, the received signal from the resonator is sent to the Rx port via a preamplifier.

This design can be further improved by additional λ/4 lines and shunt PIN diodes placed before the preamplifier. For the reverse bias voltages to be equal during Rx mode, high value resistors e.g. 20 kΩ are connected in parallel to each diode. Cross diodes can still be used at the Rx port to prevent any high voltage RF pulse leaking through the preamplifier.

A useful modification is to replace the series PIN diode at the Tx port with a shunt PIN diode configuration as shown in Fig. 17. This would eliminate the high power requirement on the PIN diode considerably and allow the designer to choose a diode with shorter carrier life time.

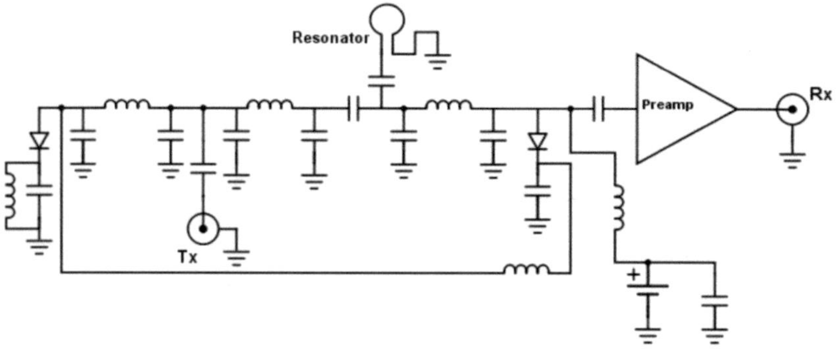

Fig. 17: Tx/Rx switch with shunt diode at the Tx side. There are three lumped element quarter wave transformers. Both PIN diodes are biased from the same source.

For quadrature driven resonators, the Tx/Rx switching can be done in two ways: 1-Tx and Rx are both in quadrature mode (Fig. 18a), 2-Tx in quadrature, Rx in linear mode by separate receive channels (Fig. 18b). In the first configuration, in Tx mode, PIN diodes are forward biased. The RF pulse applied from the Tx port is divided to two equal amplitude signals with 90° phase difference. Any reflected power at the other port of the hybrid coupler will be dissipated in the dummy 50 Ω load. In Rx mode, PIN diodes are reverse biased. The MR signal acquired from both resonators are combined and delivered to the pre-amplifier since the hybrid coupler port connected to the Tx source presents a very high impedance. In the second configuration, hybrid coupler separates only in Tx mode. In Rx mode, MR signal is received by separate Rx channels via a linear Tx/Rx switch, and the data are combined digitally during image reconstruction.

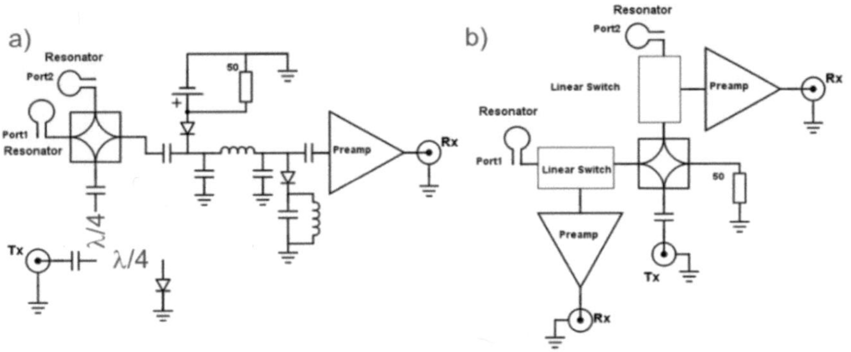

Fig. 18: Tx/Rx switch configurations for quadrature driven circularly polarized resonators. In (a), hybrid coupler divides and combines the signal both in Tx and Rx mode, whereas in (b) it is used only in Tx mode.

For inhomogeneous resonators in quadrature mode, the second configuration in Fig.18 is preferable, since the signals acquired from the regions where the B_1 fields created by the separate linearly polarized fields have different phase values. For example, if the signals acquired from both ports have a phase difference of π, they will cancel each other if they are combined by analog means without any correction.

Dedicated driver circuits apply the required DC voltage bias to the PIN diode as fast as possible. MRI systems have their own driver circuits to switch PIN diodes, and generally the DC voltage is supplied through the inner conductor of the coaxial cable connected to the resonator.

2.3 MRI of Samples with Ultra-short Transverse Relaxation Times

Conventional clinical MRI is based on acquiring signals from free or loosely bound water components in tissue. Additional data could be obtained from more rigid structures which includes bound or tightly bound water components. It was estimated that at least 20% to 30% of the hydrogen signal is not detected in clinical MRI [88].

Short T_2 of tissues can be classified into three sub-categories: porous tissue, rigid structures, and collagen containing tissue. T_2 variations in these tissues are due to molecular and nuclear phenomena, but T_2* effects include susceptibility effects, which are magnified in case of samples with porous structure. Lungs are a good example of such samples. Rigid structures such as cortical bone or dental enamel have also extremely short T_2, since the already low amount of water in these tissues is tightly bound to the rigid molecular structures that results in rapid loss of phase coherence after the RF excitation.

One third of mammalian protein mass is collagen [89]. Type-1 collagen accounts for 95% of human collagen and forms tendons, ligaments, and bones. Linearly aligned collagen tissue causes non-isotropic tumbling motion of water which destroys the coherent magnetization due to rapid accumulation of phase shifts. As a result of such rapid dephasing, collagen rich tissues have in general sub-millisecond T_2/T_2^* times. A list of short T_2 tissues along with relaxation parameters is given in Appendix B. Almost all of the tissues have more than one transverse relaxation component due to contributions from different water components. Approximate ranges for transverse relaxation components of the bound water can be listed as follows:

- loosely bound water: $T_2 \approx 1$ - 10 ms
- bound water: $T_2 \approx 100 - 400$ μs
- tightly bound water: $T_2 < 10$ μs

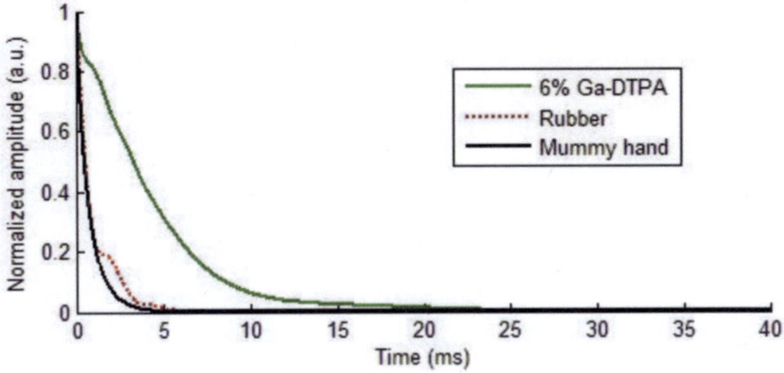

Fig. 19: Recording of the MR signal accumulated from the whole sample using a simple hard pulse excitation and immediate acquisition sequence. Signal from the mummified hand (cf. section 2.3.1.) decays faster than the 6% Ga-DTPA contrast agent solvent and a rubber sample.

With acquisition delays of about 1 ms, neither bound water nor tightly bound water is visible in clinical MRI. The ultimate purpose of this work is to make it possible to study all sources of the proton signal in the tissue using MRI and improve diagnostic capabilities of MRI. In this context, mummified samples are very useful to study improvement of MRI techniques as they have very short T_2 relaxations times. A preliminary investigation of signals from various materials with short T_2 suggest that, mummified samples are indeed extreme cases of tissue with short T_2 (Fig. 19)

In addition, MRI of mummies is challenging due to the extremely low water content.

2.3.1 An Extreme Example: Ancient Mummies and MRI in Paleopathology

Ancient mummies are unique objects of human history, and therefore non-invasive and non-destructive examination technologies are favored for paleopathology studies [90]. X-ray based methods such as x-ray radiography and computed tomography (CT) are currently the gold standard to image ancient mummies, as imaging devices are widely available, mobile X-ray units can be transported to the excavation sites, and they provide a good bone/tissue contrast [91]. Unfortunately, X-ray imaging can hardly differentiate between soft tissues. In clinical practice proton (1H) magnetic resonance imaging (MRI) is utilized for soft tissue differentiation and there have been various studies on the analysis of the water content of in vivo hard tissue such as bone, cartilage and tendons [92]–[97]; MRI of mummified tissues, however, is challenging as the MRI signal intensity is proportional to the water content which is inherently low in de-hydrated mummies. In addition, ancient mummified samples have very different physical properties than living tissue.

The MRI properties of mummified tissues are a direct consequence of the mummification process [98]. In ancient Egypt, artificial mummification was achieved by Natron, a mixture of sodium carbonate decahydrate ($Na_2CO_3 \bullet 10H_2O$), sodium bicarbonate ($NaHCO_3$) and small quantities of sodium chloride ($NaCl$) and sodium sulfate (Na_2SO_4). Natron causes excessive dehydration and is accumulated in the bone. After dehydration, the body was covered with embalming resin that was obtained from various trees and the process was completed by wrapping the skin with multiple layers of linen [99]–[103]. Advanced chemical analyses of samples of tissues, wrappings and resinous materials revealed the presence of complex mixtures of n-alkyl and cyclic biomarker components characteristic of fats, oils, beeswax, sugar gum, petroleum bitumen, ans coniferous, Pistacia and possibly cedar resins [104]–[107]. Thus, after exposing mummified bone to extreme physical conditions of temperature, pH levels and mineral accumulation over long time periods, it mainly consists of hydroxyapatite-based crystals and decreased amounts of proteins and lipids, whereas protons are primarily found in the skin and other soft tissues. In mummified samples the dominant component of the proton MR signal is formed by semi-crystalline hydroxyapatite (HAP) which is the major part of the inorganic content of the bones. Pickling, drying, freezing, and tanning prevent bone proteins from degradation but substantial collagen loss is inescapable, yet HAP crystals stay in equilibrium as long as the environmental conditions are stable [108]–[111].

Although mummified tissue has been studied with spectroscopic analysis, MR imaging has not been used in mummies since it was considered impossible due to the low water content and the short T_2* values. To overcome this limitation, in one MRI study the mummified samples were re-hydrated before MR imaging [111], [112], but this destructive method is not suitable for the valuable ancient objects. Due to the increased sodium content after mummification, ^{23}Na-MRI of an ancient mummified finger was also demonstrated, but the low signal-to-noise ratio required very long measurement times [113].

Only recently, feasibility of ^{1}H-MRI in ancient remains was demonstrated, and compared to CT, where it was concluded that even though CT remains the imaging modality of choice for ancient mummy research, MRI could provide complementary soft tissue information in some cases [114]–[116]. ^{1}H-MR images of ancient mummies in these studies showed a significantly lower quality than CT images of the same anatomical regions. The MR images had a lower SNR, and the tissue contours appeared blurred, both of which is a direct consequence of the short T_2* of the samples.

2.3.2 Hardware Considerations

During MR data acquisition, signal from short-T_2* tissues decays more rapidly, and image information encoded in the outer regions of k-space is suppressed. The signal decay thus acts as an image filter which results in a blurring of edge structures. Together with the low proton density the signal decay also reduces the available MR signal resulting in a low SNR.

To overcome the limitations, many signal averages are required to increase the SNR, and the echo time (TE) needs to be minimized. In short-TE pulse sequences TE is typically dominated by a hardware-dependent dead-time, i.e. the time delay between the end of the RF excitation and the beginning of the data acquisition (Fig. 20). The dead-time or the acquisition delay is mainly influenced by the Tx/Rx switching, ADC filtering and resonator ring-down times [117]. Here, TE is used in the literature of short TE sequences [27], [97], [116]–[121] to describe the time between the center of the RF pulse and the beginning of the acquisition (which is the time when k=0 is acquired). Thus, TE is to be interpreted as $TE = ½\, T_{pulse} + T_{delay}$.

Considering the extremely low amount of the free protons in the mummified samples, the main design criteria for the imaging resonators is to have a high sensitivity, thus to provide high image SNR. To this end, Tx/Rx resonators are recommended, so that both the Tx and Rx fields are as close to the sample as possible. A volume resonator is preferable to achieve a homogenous excitation field, which is critical in T_1 relaxation time mapping (2.3.4.3). Resonator dimensions should maximize the filling factor for the sample, so that it is possible to achieve high flip angles at lower input voltages than with commercially available resonators.

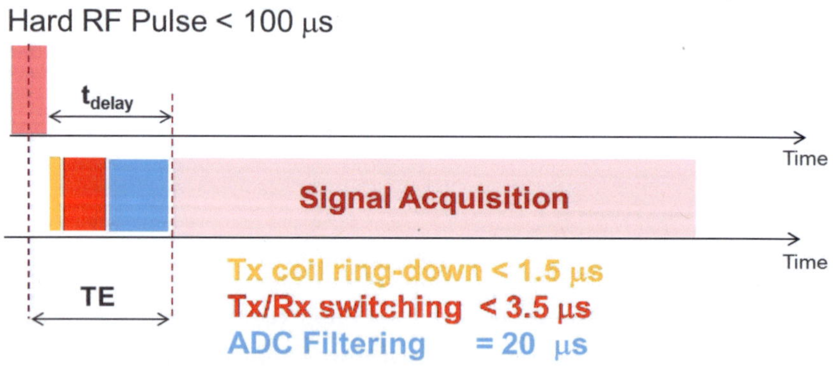

Fig. 20: Typical signal acquisition and RF pulse timing for FID based sequences. Acquisition delay depends on three factors. Design goals for each factor are given except the digital filtering which is not accessible for modifications.

2.3.3 Short-TE Pulse sequences

Recently, pulse sequences with extremely short excitation and acquisition delays have been developed to overcome signal loss due to rapid decay of transverse magnetization of spins in various tissues as, for example, in bone, cartilage, lung, or collagen fibers [27], [122]–[125]. To this end, Ultra-short echo time (UTE) pulse sequences are used with non-selective RF pulses shorter than 100 μs followed by 3D encoding gradients which apply a pure frequency encoding scheme [27]. 2D UTE approaches were also developed by using a slice selective excitation [126] and saturation based methods [127]. Alternatively, sequences such as BLAST [28], RUFIS [29], and WASPI [128] were developed where the RF pulse is applied when the gradients are switched on to further reduce echo time. Pure phase encoding techniques such as SPI [119] and SPRITE [129] also offer very short TE, yet total acquisition times are extremely long for clinical practice. PETRA combines ZTE with SPI to recover missing k-space data by SPI measurements rather than algebraic methods [120], which makes it possible to use the sequence in clinical systems that are not equipped with specially designed RF hardware, switches and ADC filters.

So far, 1H-MRI of ancient remains has been performed with the ultra-short TE (UTE) pulse sequence, which samples 3D k-space radially using a conventional frequency encoding readout [27]. In general, short-TE image acquisition strategies can be separated in two groups: frequency-encoding techniques and phase-encoding techniques. UTE is a pure frequency-encoding technique and, there-fore, it is subject to frequency domain artifacts. Single Point Imaging (SPI) is a pure phase-encoding technique which is especially useful to avoid filtering effects in k-space [119]. However, SPI has very long scan times, and it exceeds the limits of clinical gradient hardware for samples with extremely short T_2*. Pointwise Encoding Time Reduction with Radial Acquisition (PETRA) is a hybrid technique based on radial frequency-encoding of k-space with center of k-space acquired using SPI [120].

In pure frequency encoding techniques as in UTE, image information encoded in the outer regions of k-space is suppressed as a result of the rapidly-decaying MR signal. The signal decay thus acts as an image filter, which results in a blurring of the edge structures. Single Point Imaging (SPI) is an MRI technique with pure phase encoding that avoids the described filtering effect, because each point in k-space is acquired at a fixed time during the acquisition.

2.3.3.1 Ultra-short Echo Time (UTE)

UTE sequence was first used in lung imaging [27] as a projection reconstruction sequence and the intention was to reduce TE by acquiring FID signal right after the excitation. Here, although there is no echo signal but FID, TE is defined by convention as the time interval from the center of hard RF pulse to the beginning of the acquisition (Fig. 21a).

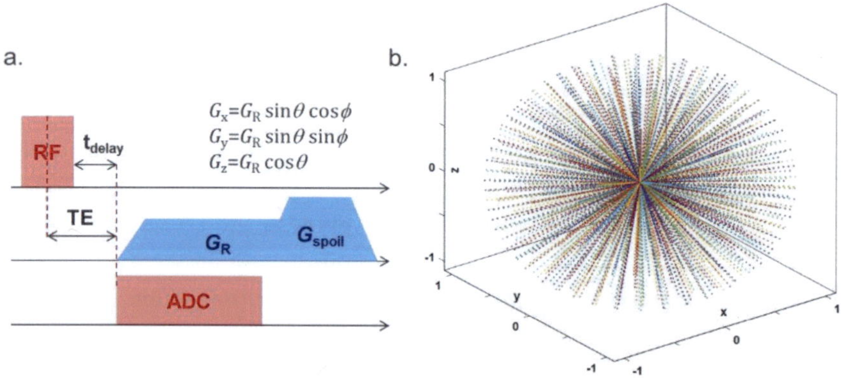

Fig. 21: UTE sequence diagram and radial trajectory. Pulse sequence diagram for UTE sequence with data acquisition starting at the ramp-up of gradients (a). The resulting k-space sampling points as radial inside-out spokes where the last data points of each spoke forms equidistant points distributed on a sphere (b).

k-space is covered in radial fashion and therefore the data is on a non-Cartesian grid: dense in the center, sparse at the periphery which necessitates application of re-gridding prior to FFT in reconstruction (cf. section 2.1.6). The strategy for signal acquisition is a pure frequency encoding since all the gradients are used as frequency encoding gradients navigating throughout the k-space in radial inside/out fashion (Fig. 21b). Data acquisition starts at the ramp-up of frequency encoding (readout) gradients, which results in even more dense sampling at the center.

In 3D non-selective excitation is achieved since no gradients are present during the RF pulse. 2D UTE is implemented in two ways:

- applying a slice selection gradient during a half sinc pulse [126], [130]
- creating a saturation band and digitally subtract two non-selective UTE acquisitions with and without selective saturation of the slice of interest [127]

2D UTE pulse sequences can be used as localizer for samples with extremely short T_2 values.

In UTE sequences blurring is the major problem due to the filtering effect from center to the radius in k-space due to rapid T_2^* decay. This problem is common to all frequency encoding based techniques.

2.3.3.2 Single Point Imaging (SPI)

In SPI (Fig. 22), only a single k-space data point is acquired per TR, and 3D k-space is sampled directly on a Cartesian grid [119]. Therefore, SPI requires rapidly switching gradients in order to encode information efficiently in the outer regions of k-space.

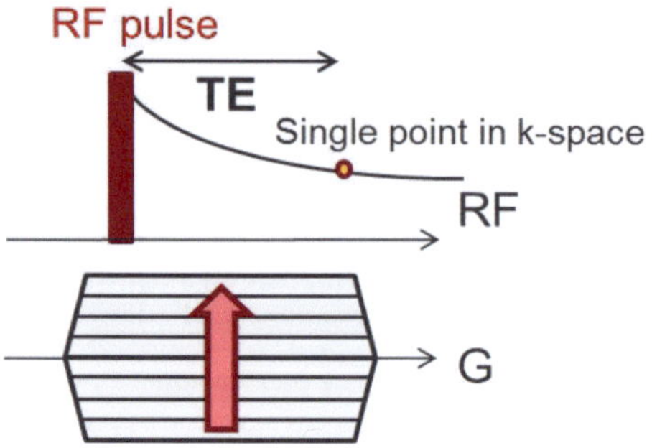

Fig. 22: SPI pulse sequence. Gradients are active during a very short block RF pulse. Here minimum echo time depends on the furthermost k-space point which is limited by the gradient strength.

An alternative approach to the gradient-switching is to employ continuously ramped gradients, which is a more common way of implementing SPI [129]. In this case, SPI requires long and strong gradients, which results in excessive heating of the gradient coils. Gradient heating can affect the gradient fields [131], and detailed heating models have been proposed in literature to assess this problem in detail [132], [133]. In addition to the heating, unwanted slice selectivity is imposed in the images, because the gradients are active during the RF excitation. Post processing schemes have been proposed for eliminating the slice selection artefacts [134].

2.3.3.3 Pointwise Encoding Time Reduction with Radial Acquisition (PETRA)

In PETRA, an RF excitation is followed immediately by a radial k-space acquisition, but the encoding gradient is already switched on during the RF pulse to minimize the acquisition delay as in Zero Echo Time (ZTE) method [135]. However, since there is a time delay between the RF pulse and the start of the acquisition, the center of the k-space is not acquired. k-space center is later filled by a single point acquisition method (cf. above) as shown in Fig. 23 as a 2D representation. Thus PETRA is a hybrid between UTE and SPI.

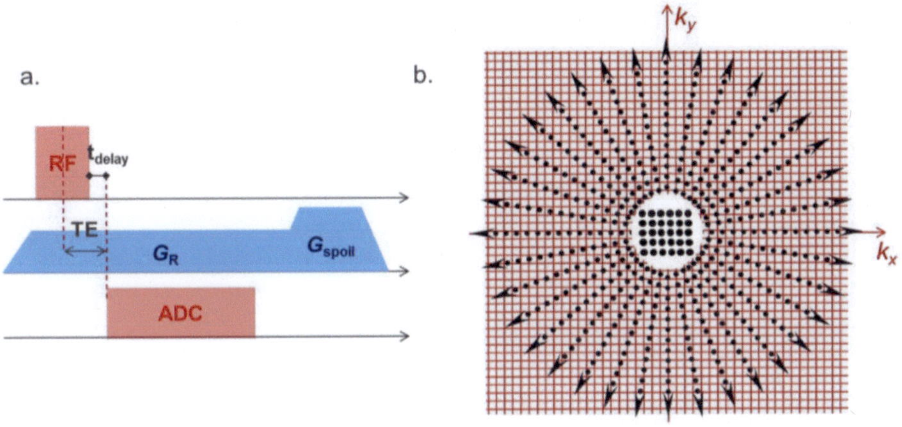

Fig. 23: PETRA sequence diagram. Gradients are always on during the RF pulse for the radial acquisition of outer sections of k-space. The center of k-space is missing because of the acquisition delay between the end of RF pulse and the beginning of the acquisition (a). SPI is used to cover the missing data points. PETRA k-space locations on a representative 2D grid (b).

2.3.4 Parametric Mapping and Data Analysis

Quantitative analysis is done by SNR comparison and relaxation time mapping.

2.3.4.1 Signal to Noise Ratio Measurement

SNR is calculated pixelwise as the ratio of image intensity on a region of interest (ROI) and the standard deviation of a background region where no signal present [136]. Pulse sequence parameters have direct influence on the image SNR. Therefore, to eliminate these effects, SNR scaling is done by scaling the image intensity values for different pulse sequences with respect to the imaging parameters: sampling time/bandwidth, total acquisition time, and the repetition time [137], [138].

$$SNR = \psi\, V \sqrt{T_s}\, \sqrt{\frac{T_{image}}{TR}} \sin(\propto) \left(\frac{1 - e^{-T_R/T_1}}{1-\cos(\propto)\cdot e^{-T_R/T_1}}\right) e^{-T_E/T_2^*} \qquad (2.3.1)$$

Here, ψ is the normalized system SNR, V the voxel volume, T_s the sampling time, T_{image} the total imaging time, T_R the repetition time, T_E the echo time (i.e. acquisition delay or dead time) and α the flip angle..

2.3.4.2 T_2* Relaxation Mapping

For samples with extremely short T_2, UTE sequence can be used for T_2* quantification. Keeping all imaging parameters constant, TE (or acquisition delay) is changed and images are acquired for a range of TE values. For each pixel or group of pixels from a selected RO, the square of the intensity of each data points are then fit to a mono-exponential decay function with a time constant of $2/T_2$*, plus an arbitrary time constant representing the noise. An example of fitted curve for range of TE values from 50 μs to 2.3 ms is shown in Fig. 24.

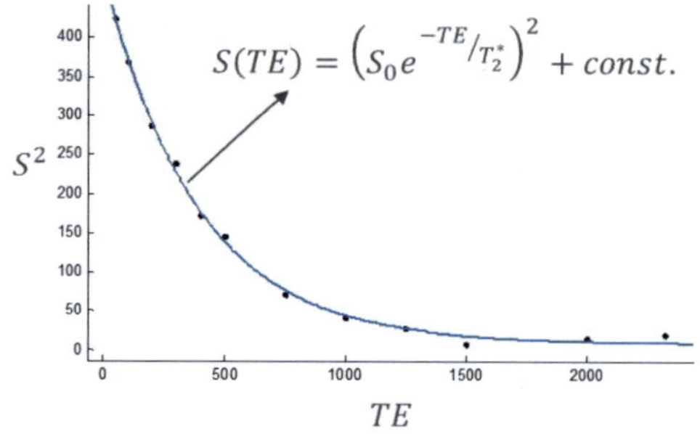

Fig. 24: T_2* calculation for a strip band using an exponential fit over the data points acquired at different TE. Resulting T_2* value is 205 μs.

2.3.4.3 Modified FLASH Equation for T_1 Relaxation Mapping

The method for estimation of T_1 re-writes Eq. 2.1.25 for a curve fitting for a range of TRs [139]:

$$\frac{M\alpha}{\sin\alpha} = e^{-TR/T_1}\frac{M\alpha}{\tan\alpha} + M_0(1 - e^{-TR/T_1}) \qquad (2.3.2)$$

This equation is a linear plot of $M_\alpha/\sin\alpha$ against $M_\alpha/\tan\alpha$ with a slope of exponential T_1 decay. For T_1 estimation, images for a range absolute flip angles are obtained and pixel-wise curve fitting is applied. The sensitivity of this method is sufficient for most of the applications, which can be calculated by differentiation of eqn.2.3.2 with respect to $1/T_1$:

$$\frac{\partial M_\alpha}{\partial(1/T_1)} = \frac{M_0 TR e^{-\frac{TR}{T_1}} \sin\alpha(1-\cos\alpha)}{\left(1-\cos\alpha e^{-\frac{TR}{T_1}}\right)^2} \tag{2.3.3}$$

$$\frac{\partial}{\partial\alpha}\left(\frac{\partial M_\alpha}{\partial(1/T_1)}\right) = 0 \tag{2.3.4}$$

$$\cos\alpha = \frac{2e^{-\frac{TR}{T_1}}-1}{2-e^{-\frac{TR}{T_1}}} \tag{2.3.5}$$

Eq. 2.3.5 can be interpreted that for the range of α over 30° - 100° in steps of 10°, T_1 values between TR-20TR can be estimated with reasonable accuracy.

In case of extremely rapid T_2 decay, however, the signal loss during RF excitation pulse should be taken into account as shown in the modified FLASH equation [140] below:

$$M_\alpha = M_0 \frac{(1-e^{-TR/T_1})k_1}{1-k_2 e^{-TR/T_1}} e^{-(\frac{TE}{T_2^*}-\tau)} \tag{2.3.6}$$

for

$$k_1 = e^{-\tau}\frac{\alpha}{\varphi}\sin\varphi \;;\; k_2 = e^{-\tau}\left(\cos\varphi + \frac{\tau}{\varphi}\sin\varphi\right) \;;\; \varphi = \sqrt{\alpha^2-\tau^2} \;;\; \tau = \frac{T_{RF}}{2T_2^*}$$

To sum up, for a full relaxation time mapping of samples with extremely short T_2, the workflow starts from flip angle mapping, then T_2^* mapping, and finally T_1 mapping is done using the modified FLASH equation.

2.4 Concurrent Acquisition and Excitation (CEA)

The main challenge in CEA is the very high amount of isolation required between the Tx and Rx circuits. In concurrent operation of Tx and Rx, the MR signal level is typically 100 dB smaller than the RF radiation coupled from Tx to Rx. To detect MR signal with a reasonably high SNR, the Tx induced signal in Rx should be reduced down the Tx noise floor as demonstrated in Fig. 25 for a typical MR signal size and respective Tx RF pulse. In his initial setup, Bloch applied geometric decoupling [4], where, separate Tx and Rx resonators were placed orthogonal to each other around the imaging volume. With this orthogonal setup, the cross-talk between Tx and Rx resonator is minimized so that the unwanted RF excitation signal is highly suppressed during data acquisition. More advanced techniques proposed for isolation of Tx and Rx units were described in section 2.1.5.2. In this section, two novel methods, *active decoupling* and *dynamic analog cancellation* will be introduced.

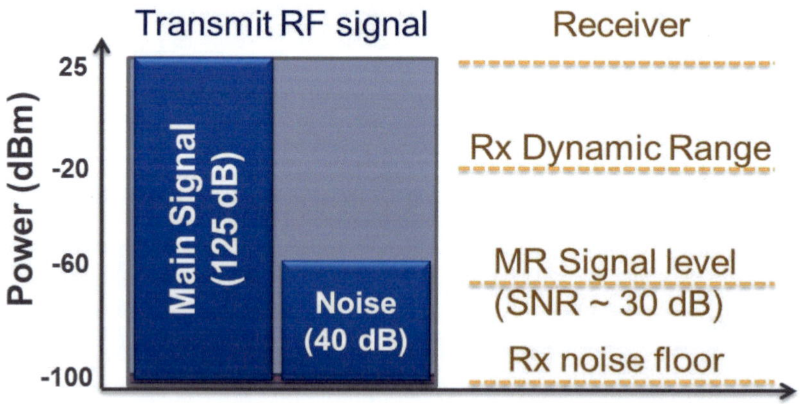

Fig. 25: Representative signal and noise level comparison.

2.4.1 Active Decoupling of RF Resonators Using a Parallel Transmit Array System

Active decoupling method uses multiple Tx elements for cancellation of the Tx-induced RF signal in the Rx resonator and demonstrated direct detection of CEA MR signal in a 3 T clinical MRI system [52].

It was recently shown that a parallel transmit array system can be used to cancel RF currents that are induced on a bare wire [141]. In this work this concept is adapted to cancel the RF currents that are induced on a loop receive resonator. In general, the combined RF field generated by a set of Tx resonators, induces a current in a receive resonator, which is observed at the output of the receive resonator as a B_1-induced voltage. Using an additional set of Tx resonators (decoupling resonators), it is possible to cancel the B_1-induced currents in the Rx resonator by adjusting their RF phases and amplitudes. For a transmit array system with N resonators, some of which are used as Tx resonators and some as decoupling resonators, the total current I induced in a Rx resonator can be expressed as:

$$I = \sum_{n=1}^{N} a_n I_n = 0 \tag{2.4.1}$$

Here, I_n is the B_1-induced current, and a_n is an arbitrary complex coupling coefficient for the nth Tx resonator, which represents the geometric and electromagnetic factors that affect the amount of coupling between the Rx resonator and the nth Tx resonator.

In its simplest form, the phase and amplitude (PA) decoupling system can be implemented with two Tx resonators – one to apply the RF excitation, the other for decoupling. By adjusting the amplitude and phase a_2 of the decoupling resonator, one can theoretically achieve a cancellation of the total current in the receive resonator: $a_1 I_1 + a_2 I_2 = 0$. In practice, it is sufficient to measure the voltages that are induced by each of the Tx resonators individually, and to solve

$$V_1 + cV_2e^{j\theta} = 0 \quad (2.4.2)$$

for the two parameters c and θ.

In CEA, Tx stability during the measurement is crucial, because a residual leakage signal is also acquired together with the MR signal due to limited isolation. Other CEA methods (section 2.1.5.2.) used a theoretical function for digitally cancelling out the residual Tx leakage signal, which failed in cases of motion or changes in resonator loading with the existing decoupling schemes. In [142] and [143], *in vivo* CEA MRI was also demonstrated using pick-up coils (PUCs) to monitor Tx signal during excitation.

2.4.2 Automated Dynamic Analog Cancellation using Dynamic Feedback from MRI System

The isolation of Tx and Rx units for simultaneous performance of both is a problem in telecommunication electronics to benefit from the full bandwidth in data transfer, and this type of communication is called 'Full Duplex'. To comply with MRI, we focus on a specific type of full duplex communication: In band full duplex (IBFD), since in MRI the RF excitation and the signal from the spins are within the same frequency band. Methods for IBFD can be addressed in three main sections: passive and active suppression, and digital cancellation [144]. Passive suppression methods include antenna separation [145], antenna decoupling [146], and circulator isolation [147]. Active suppression methods are based on employing actively controlled circuits to subtract a copy of the Tx signal from the Tx induced leakage signal in Rx resonator. This can be implemented either at RF [148] or in baseband. Recently Bharadia et al. suggested a combined active analog and digital cancellation which can operate in real-time [149]. Following a similar approach in analog cancellation as in [149], very high amounts of on-resonant decoupling were reported in [150] to implement CEA MRI. A similar work which is also based on the work described in [149] demonstrates *in vivo* MRI with CEA [151].

2.4.2.1 Analog cancellation circuit elements

Analog cancellation circuit composes of voltage controlled phase and amplitude regulation elements. Working principles of these elements are described below.

2.4.2.1.1 Hybrid coupler based variable phase shifter

A phase shifter is a two-port RF network device that changes the phase of an input signal. Phase shifting circuits are used in phased antenna arrays, radar arrays and other communications systems. Electrically, there are many methods to implement such a shift in phase without modifying other characteristics of the input signal. Mechanical switching between various transmission lines, and FET based active phase shifters are few of them. Hybrid coupler based phase shifters are also commonly used [152].

A schematic of a hybrid coupler with arbitrary reflections at two ports is given in Fig .26. A normal function of hybrid coupler when all ports are matched to 50 Ω is the following: the incident power at port 1 (input port) is divided approximately equally (3 dB division) between ports 2 and 3 (direct and coupled ports, respectively), with very little power reaching port 4 (isolated port) [153]. However, the incident and reflected voltages for an arbitrary reflection coefficient writes:

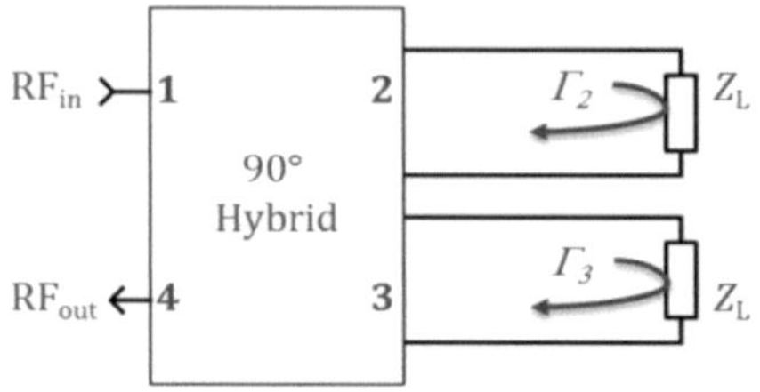

Fig. 26: A four port 90° hybrid coupler with arbitrary reflections at ports 2 and 3.

$$V_1^- = \frac{-j}{\sqrt{2}}V_2^+ + \frac{-1}{\sqrt{2}}V_3^+ \tag{2.4.3a}$$

$$V_2^- = \Gamma_2 \frac{-j}{\sqrt{2}}V_1^+ \tag{2.4.3b}$$

$$V_3^- = \Gamma_3 \frac{-1}{\sqrt{2}}V_1^+ \tag{2.4.3c}$$

$$V_4^- = \frac{-1}{\sqrt{2}}V_2^+ + \frac{-j}{\sqrt{2}}V_3^+ \tag{2.4.3d}$$

A simple algebraic operation leads us to the following relation:

$$V_4^- = j\Gamma V_1^+ \; ; \; \Gamma = |\Gamma|e^{-j\phi} \tag{2.4.4}$$

We can replace the arbitrary impedance Z_L with a variable capacitor connected to ground. Relation between the reflection coefficient and a capacitor connected to ground can be calculated as follows:

$$\Gamma = \frac{Z_C-50}{Z_C+50} = \frac{(jwC)^{-1}-50}{(jwC)^{-1}+50} = e^{j\phi} \tag{2.4.5}$$

$$\phi = -2\tan^{-1}(50C\omega) \tag{2.4.6}$$

The reflection has unitary amplitude. Additional phase shift tends to 0 or π for zero or infinite capacitance, respectively. Therefore using a voltage controlled capacitor i.e. varactor, voltage controlled phase shifter can be implemented. The powers reflected from ports 2 and 3 have phase angles governed by the capacitance of the varactor diode. Because of the hybrid coupler's action in reverse, the reflected powers reaching port 1 almost cancel each other. The reflected powers reaching port 4 add since they are in phase. With a careful selection of the capacitance range of the varactor, the range of phase shift and flatness over the frequency range of interest can be optimized.

2.4.2.1.2 Variable attenuators

Pin-diode based attenuator

A resistive π-network functions as an attenuator. In Fig. 27, shunt resistors R_1 and the series resistor R_3 (Eq. 2.4.7) are set in a π-network to achieve an attenuation A = 20 log(K) while simultaneously providing an input and output impedance which matches the characteristic impedance of the system:

$$R_1 = Z_0 \frac{K+1}{K-1} \; ; \; R_3 = Z_0 \frac{K^2+1}{2K} \tag{2.4.7}$$

where, K is a constant to determine the required attenuation, and Z_0 is the characteristic impedance of the system.

A PIN diode can be used as a current-controlled variable resistor. Series resistance of a diode is inversely proportional to the applied forward current. Three diodes can be used to replace the fixed resistors of the π-network to create a voltage controlled variable attenuator. In this configuration, the diodes that replace the resistors labelled R_1 share a single DC bias network; whereas the diode that replaces R_3 is driven by a separate DC bias source, which is preferably a variable DC voltage source.

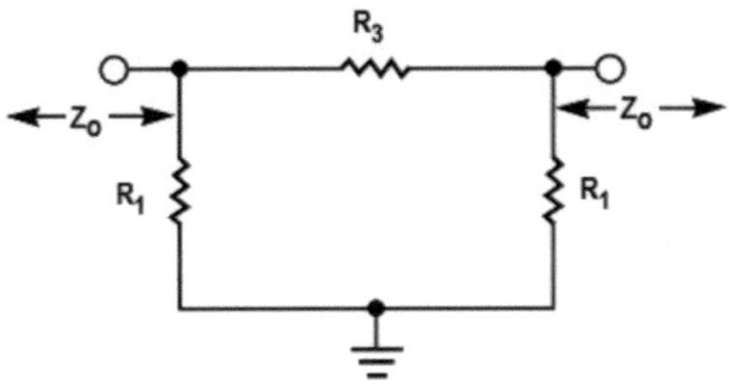

Fig. 27: A resistive π network applies controlled attenuation to the input RF signal.

Hybrid coupler based attenuators

An alternative design for a variable voltage controlled attenuator is a reflection type design using a hybrid coupler as in the phase shifter, but with PIN diodes rather than varactors. For a 50 Ω quadrature Hybrid, the attenuation A is a function of the diode resistance, R_S.

$$A = 20\,log\,(1\,/\,(1 + 2R_S/Z_0))\ [\mathrm{dB}] \tag{2.4.8}$$

The quadrature hybrid configuration can control twice the power of the simple series or shunt diode attenuators because the incident power is divided into two paths by the hybrid. The maximum power dissipated in each diode is only 25% of the total incident power and this occurs at the 6 dB value of attenuation. Branch load resistors in parallel to the PIN diodes are used to make the attenuator less sensitive to differences between individual diodes and to increase the attenuator power handling by 3 dB. The branch load resistors must be able to dissipate 50% of the total incident power at maximum attenuation.

2.4.3 Data Processing in CEA

As discussed in section 2.1.5, FM pulses are used for CW NMR or rapid-scan correlation experiments. In response to such frequency modulated pulse (e.g. chirp pulse with linear frequency modulation), the phase of the resultant transverse magnetization, M_{xy} is a quadratic function of resonance offset. Such quadratic phase dependence is not favored in conventional MRI applications, and it can even cause signal loss or artifacts in the image as it cannot be refocused with linear gradients when used for slice selection. In CEA, the quadratic phases are not a problem as signal is acquired concurrently.

The rapid passage operating in the linear region creates a highly linear spin response for flip angles up to 90° [154] and the quadratic phase of the resultant M_{xy} is removed with a cross-correlation method identical to that used to recover phase information in stochastic NMR spectroscopy [155]. Accordingly, the spin system response can be treated as a linear system, in which case the response $s(t)$ to an input function $b(t)$ is the convolution process given by

$$s(t) = h(t) \otimes b(t) \tag{2.4.9}$$

where $h(t)$ is the unit impulse response i.e., FID. The Fourier transform of $h(t)$, $H(\omega)$, is the unit response function i.e. a projection of the object in frequency domain [41]. According to Fourier theory, the convolution in the time domain is a complex multiplication in the frequency co-domain:

$$S(\omega) = H(\omega)B(\omega) \tag{2.4.10}$$

for

$$b(t) = \omega_1(t)e^{-i\int_0^t(\omega_{RF}(\tau)-\omega_c)d\tau} \tag{2.4.11}$$

where, ω_c is the center frequency in the sweep range. A projection of the object is obtained by correlation, which is a complex conjugate multiplication by the pulse function in the frequency domain [37]:

$$H(\omega) = \frac{S(\omega)B^*(\omega)}{|B(\omega)|^2} \tag{2.4.12}$$

In a CEA experiment, the acquired MR signal, $s_{Rx}(t)$, is written as an additive combination of the convolution signal and a frequency-dependent leakage component $\vartheta_{leak}(t)$, which is the remaining B_1-induced voltage due to imperfect decoupling:

$$s_{Rx}(t) = h(t) \otimes b(t) + u_{leak}(t) \tag{2.4.13}$$

and we can write the chirp signal as:

$$b(t) = b_0 e^{-i\pi\frac{f_s}{t_{acq}}t^2} = b_0 e^{-i\omega(t)\mathrm{t}} \tag{2.4.14}$$

The leakage can be determined experimentally by measuring with a coupled pick-up coil or based on fitting to the theoretical formula in Eq. 2.4.13. As the frequency is swept linearly through time, $\omega(t) = \pi\left(f_s/t_{acq}\right)$t, the leakage response can also be treated in the frequency domain, which is an intrinsic property of the chirp signal. The ratio of the frequency sweep range f_s and the total RF duration t_{acq} is regarded as the frequency sweep rate.

2.4.4 CEA in Steady State: T1-Modulation Effects of Frequency Sweep

In MRI with CEA, frequency modulated radiofrequency (RF) excitation pulses are applied. Thus, magnetization across the sample is excited sequentially by the combined action of the magnetic field gradients and the RF pulse. Gradient directions change during data acquisition which results in TR variation across the sample (Fig. 28), leading to changes in transverse magnetization. This signal modulation causes a non-ideal point spread function (PSF), and is a potential source of artifacts.

Investigation of the T_1 modulation effects helps to find an optimal approach in designing pulse sequences for CEA.

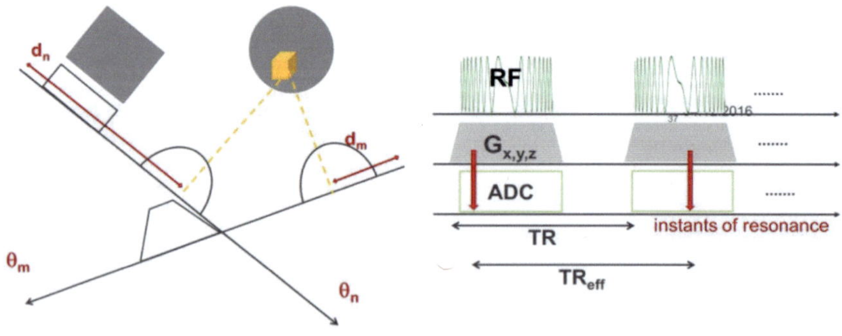

Fig. 28: Demonstration of the frequency sweep during readout gradients resulting in different instants of resonance for two different radial spokes for an off-isocenter voxel. The actual TR between two moments of excitation is referred to as effective TR, TR_{eff}.

2.5 Electroglottography (EGG) Measurements during Dynamic Lung MRI

EGG is an indirect method of determining the relative vocal fold (Fig. 29) contact area during phonation. The vocal cords are composed of three layers: the vocal muscle is the main body of the vocal fold, superficially lays the ligament as intermediate and deep part of the lamina propria and the Reinke's space as superficial layer containing loose fibrous components. The vocal fold is covered with epithelium [156]. The muscle tension, the tissue stiffness and fiber components determine the characteristics of the vocal fold vibration [157]. EGG is commonly applied in medical studies of the vocal system and for pedagogical research on singing [158]–[161]. An EGG device measures the impedance change in the larynx during vocal fold activity by applying a weak alternating current through two electrodes on the surface of the neck. The electrical impedance varies slightly with the area of contact between the moist vocal folds during phonation [162]. Therefore, the shape of the EGG wave form can be used to infer about the vocal fold contact patterns [163]. This technology is recommended to be part of the evaluation of patients vocal fold vibration in clinical context [164]. The glottis vibrates at acoustic frequencies from tens of Hz to a few kHz [165]. To overcome the high impedance of the poorly conductive outer layer of the skin and the myelin insulation of muscle fibers between the electrodes and the glottis, EGG devices use high-frequency electrical currents of several MHz. After demodulation, typical information extracted from the EGG waveforms include the fundamental frequency, the contact quotient (duty cycle), and regularity parameters such as jitter and shimmer i.e. temporal change of frequency and amplitude of the vocal fold vibration, respectively.

Dynamic MRI has recently been shown to be a promising tool in speech and phonation analysis [166]. MRI can provide dynamic and static information about the oropharyngeal cavity, hypopharynx and larynx during phonation, but it is also able to measure pulmonary and diaphragm movements as well as pulmonary volume [167]–[170]. For dynamic imaging of the vocal tract, fast two- and three-dimensional MRI techniques are used with a temporal resolution between 33 ms and 125 ms [171]–[179]. The MRI measurement of lung volume and diaphragm motion during singing is of particular interest in studies of the breathing apparatus and breathing strategies in both trained and untrained voices. Since the human voice production is based on the onset of the subglottic pressure, i.e. the pressure below the vocal folds [157], separate evaluation of the expiratory forces produced by the lungs is not meaningful without evaluation of the area where the force acts (the vocal folds). Since the latter system is changing in the time domain it is of great importance to include measurements of vocal fold oscillations (such as EGG signals) during measurements of the breathing apparatus.

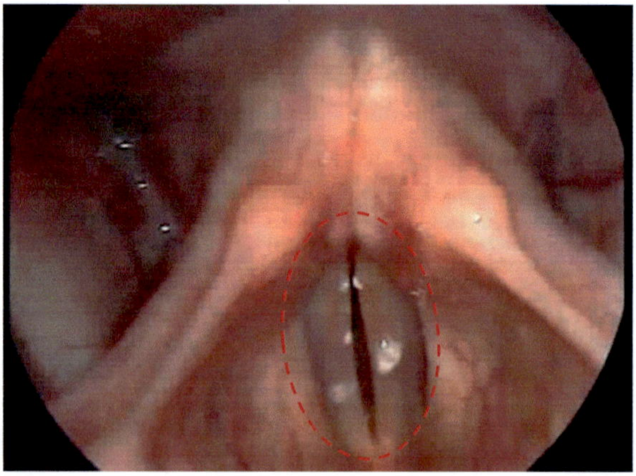

Fig. 29: An image from a video captured with a stroboscope endoscopy of a male patient. Endoscope was placed inside the throat.

3 Methods

The following three chapters describe the different concepts introduced in this work, starting with optimization of state-of-the-art MRI pulse sequences with ultra-short TE for imaging of ancient mummies [180] in chapter 3.1. In chapter 3.2, two novel methods for implementing CEA in clinical MRI systems are described. Feasibilities of active decoupling method [142], which provides isolation over 70 dB between Tx and Rx resonators, and then dynamic analog cancellation method, which increases the amount of isolation up to 90 dB and provides automated decoupling, were demonstrated in a clinical MRI system. Chapter 3.3 introduces simultaneous functional imaging of lungs and electroglottography (EGG) measurements. For the first time, EGG was measured during MRI to analyze breathing dynamics and vocal fold function concurrently [181].

3.1 MRI of Samples with Ultra-short Transverse Relaxation Times

The challenges of MRI of samples with ultra-short T_2 were discussed in chapter 2.3. In this chapter, the techniques to overcome those challenges and experimental details will be described. These techniques and materials are partly explained in [180].

3.1.1 Ancient Mummified Human Hand from Egypt

To compare the UTE, SPI, and PETRA pulse sequences, an artificially embalmed ancient Egyptian mummified left hand (ca. 1500 – 1100 BC), which is stored at room temperature, was used (Fig. 30). The hand is of unknown provenance and it stems from the former collection of Musée d'Orbe, Orbe,

Switzerland. It was donated to Prof. Dr. Frank Rühli (head of Institute of Evolutionary Medicine, University of Zurich).

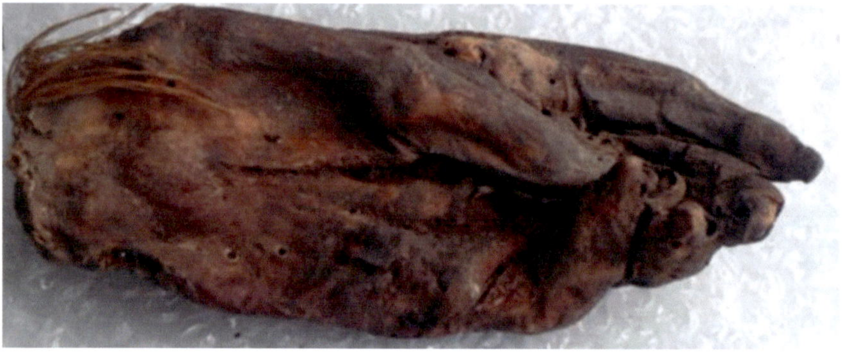

Fig. 30: A photo of the left hand of the artificially embalmed ancient Egyptian mummy.

For comparison with the MR techniques described in section 2.3.3, multidetector CT images (Philips Brilliance CT 40; Philips Healthcare, Best, the Netherlands) of the sample were obtained at the Balgrist University Hospital, Zurich, Switzerland with 0.76 mm section thickness, 512 x 512 matrix, 0.74x0.74 mm pixel size by applying 120 kV, 72 mA.

3.1.2 Modifications to the Short-TE Pulse Sequences

The pulse sequences were programmed using IDEA, the sequence development environment of Siemens. PETRA sequence was supplied by the manufacturer. Unless specified otherwise, the reconstruction of the MR images from the acquired raw data was done in MATLAB 9.0 (The MathWorks®, Inc., Natick, Massachusetts, United States) by re-gridding.

3.1.2.1 Gradient Switching Corrections in UTE Sequence

3D UTE sequence is composed of a hard RF pulse followed by data acquisition during the ramp up of the readout gradients, where a short delay of few tens of microseconds occurs in between due to the system-specific dead time. Full 3D k-space coverage is achieved by radial spokes starting from the center of k-space, and the end tip of all the spokes form equidistant points on a sphere. (cf. section 2.1.6).

In this study, for imaging the mummy hand, a rectangular RF pulse with a duration of 60 μs was applied, and the radial data acquisition started 40 μs after the end of the pulse yielding a TE of 70 μs. Other imaging parameters were as follows: number of radial spokes = 50000; acquisition bandwidth = 639 Hz/pixel; TR = 10 ms; $\alpha = 10^{\circ}$; FOV = $(206\ \text{mm})^3$; isotropic voxel size = $(0.9\ \text{mm})^3$; total scan time = 10 min. Data acquisition started during the ramp-up of the radial readout gradient. For image reconstruction, k-space data were gridded onto a Cartesian grid followed by a 3D fast Fourier transform using a manufacturer-provided prototype reconstruction algorithm.

Although the algorithm provided by the manufacturer performs well with standard routines, it had the following disadvantages: it was not applicable to all the MRI systems, the software was specific to a certain system version, and it prevented us from implementing new techniques in reconstruction. Therefore, a re-gridding reconstruction algorithm was implemented in Matlab for Siemens raw data, independent of the MRI system version (Appendix H). Within the reconstruction pipeline, algorithms were implemented to correct for gradient waveform based errors.

A major problem of a radial k-space trajectory is its sensitivity to inaccuracies in gradient timing. Since the direction of the radial spokes changes for each repetition, small delays in gradient waveform due to system imperfections result in inconsistencies of k-space sample locations and consequently, degradation of the image quality.

For correction of the gradient delay based errors, a method was used which did not necessitate extra scan time or pre-emphasis gradients. All the necessary information was extracted from the acquired imaging data itself. In the acquired 3D radial k-space data set, radial spokes in +x and −x directions were used to calculate the shift in k-space sampling locations due to the delay of the x gradient with a cross-correlation analysis [182]. The same calculation was repeated for y and z directions, and the resulting shifts were weighted with respect to the radial spoke direction. The cross-correlation function of opposite spokes has a peak at the shift distance (Fig. 31a). For a more precise estimation, the shift was estimated from the slope of the signal's phase using a linear curve fit to the signal portion with a reasonable magnitude (Fig. 31b). In Appendix C, the developed Matlab function script for estimation of gradient delays in radial MRI raw data is listed.

To test the effect of correction for the gradient delays, a custom-built resolution phantom was used. A reference GRE image slice is compared to a radial UTE image of the identical slice. Imaging parameters for GRE were FOV = 280 mm, TR = 20 ms, TE = 2.04 ms, α = 25°, 395 Hz/Px bandwidth, 0.7 mm resolution and 2 mm slice thickness. For UTE, the same parameters as in section 3.1.1 were used.

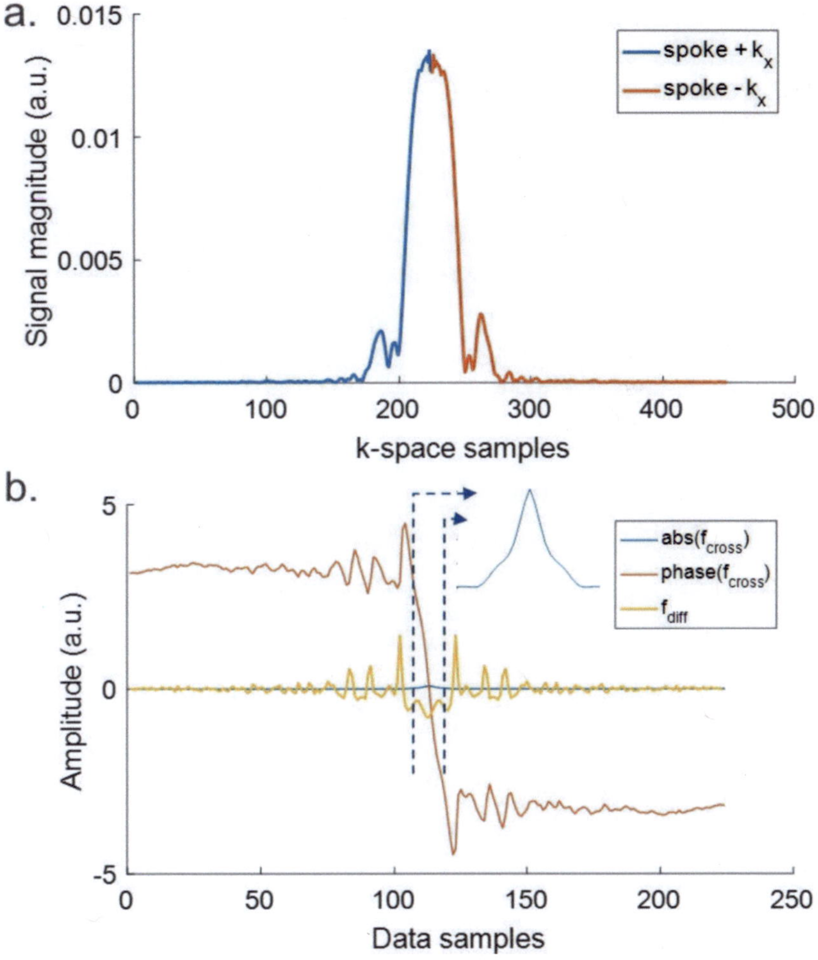

Fig. 31: Cross-correlation between spokes with opposite directions was used to calculate the displacement in k-space due to gradient timing errors. a. Opposite spokes plotted together. b. Steps of gradient delay calculation using the slope of the phase of the cross correlation function within a reasonable support. Here f_{cross} is the cross correlation function. Amplitude of f_{cross} is used to find the appropriate support region with reasonable SNR. f_{diff} is the derivative of the unwrapped phase of f_{cross}.

3.1.2.2 Gradient-friendly Modifications in SPI

A saddle-like k-space ordering for SPI sequence with continuous gradient-ramp was introduced, which avoids keeping the gradients active at high levels for long times. To this end, instead of the conventional monotonical sampling orderings in k-space, the direction of the gradient is changed quickly, and the gradients are not kept on at a high level for long time. Steep jumps in k-space are also avoided to maintain continuous gradient-ramp feature of SPI (or SPRITE). Samples of the k-space were ordered such that the 3D k-space was divided into cubic shells and each shell was travelled following contours in a saddle-like trajectory (Fig. 32e,f). It was shown that with this k-space ordering, gradient coil heating is reduced [183].

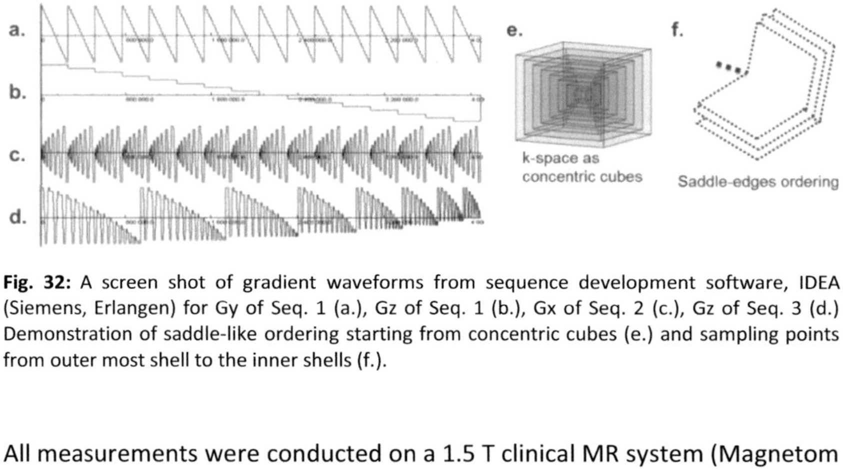

Fig. 32: A screen shot of gradient waveforms from sequence development software, IDEA (Siemens, Erlangen) for Gy of Seq. 1 (a.), Gz of Seq. 1 (b.), Gx of Seq. 2 (c.), Gz of Seq. 3 (d.) Demonstration of saddle-like ordering starting from concentric cubes (e.) and sampling points from outer most shell to the inner shells (f.).

All measurements were conducted on a 1.5 T clinical MR system (Magnetom Tim Symphony, Siemens AG, Erlangen). Three SPI sequences with continuously ramped gradients were designed: standard step-wise k-space ordering (Seq. 1), rectangular-spiral k-space ordering (Seq. 2), and saddle-like k-space ordering (Seq. 3). In Seq. 2, a rectangular-spiral trajectory was applied for each 2D slice to cover a 3D k-space. In Seq. 3, Gz of the saddle-like trajectory (Fig. 32d). Gradient waveforms that showed significant difference are given in Fig. 32a-d. Temperature information was obtained from sensors placed in gradient coil connectors. Heating was tested with sequence parameters TR = 6.4 ms; effective TE = 0.3 ms; 64x64x64 samples; readout FOV = 240 mm.

In addition, to minimize unwanted slice selection in SPI with continuous gradient-ramp, a novel approach in sequence design was described, where the gradient level is lowered during RF excitation without switching the gradients off (Fig. 33).

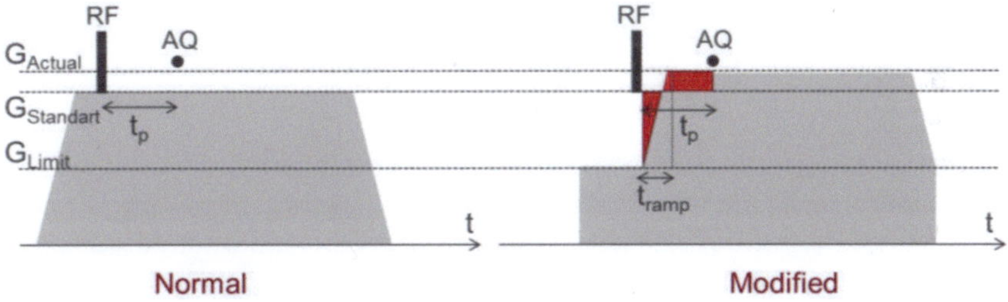

Fig. 33: Limitation of gradient level on RF excitation, $\boldsymbol{G_{Limit}}$. $\boldsymbol{G_{Actual}}$ is the new peak amplitude for the modified gradient shape.

The gradient level amplitude during readout is calculated as follows:

$$G_{Actual} = \frac{G_{Standard} \cdot t_p - G_{Limit} \cdot t_p/2}{t_p - t_{ramp}/2} \tag{3.1.1}$$

where $G_{Standard}$ is the gradient level before modification, t_p is acquisition delay after RF excitation, G_{Limit} is the limited gradient level and t_{ramp} is the ramp duration. Note that the effective TE increased due to the limited slew rate. Bloch equation simulations were performed for slice profile calculation at the maximum field strength of the sequence (Fig. 34).

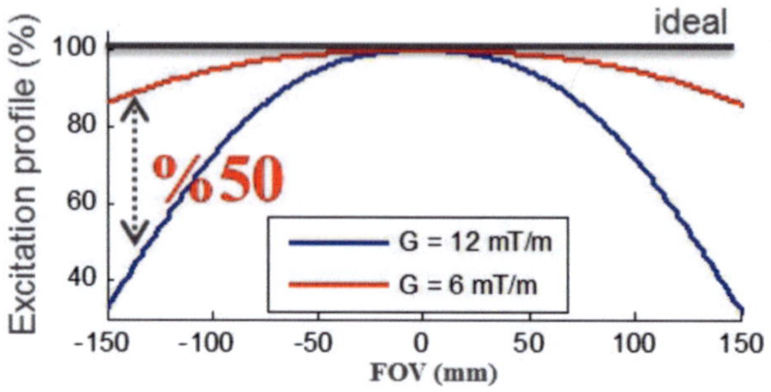

Fig. 34: Excitation profile simulation of sequences with different gradient levels. When the gradient level was halved during the RF pulse, a smoother slice was excited.

To test the effect of the gradient amplitude modulation, imaging experiments were performed with TR = 2 ms, FOV = 340 mm, 64x64x64 samples, effective TE = 0.3 ms, α = 1° using a Rx-only loop resonator (Ø 25 cm) and a rubber phantom (T_2* = 600 μs). SPI images with regular and modulated gradient strength were reconstructed for comparison.

Both methods were demonstrated for SPI with continuous gradient-ramp in a 1.5 T clinical MRI system and using a rubber phantom with a short T_2* value. Without loss of generality, the gradient modifications are applicable to other techniques, for which a pure phase encoding or a continuous gradient-ramp is employed.

In mummy imaging experiments, a rectangular RF pulse is applied, and the k-space encoding is done with a gradient with variable amplitude, which is switched on before the RF pulse of 10 μs duration to minimize the acquisition delay. The rapid gradient switching was avoided by continuous gradient-ramps as in SPRITE [129]. At 1.5 T, SPI was performed with a voxel size of 1.7x1.7x3.4 mm^3 due to gradient amplitude limitations (G_{max} = 30 mT/m). Other imaging parameters were as follows: FOV = $(212\ mm)^3$, $\alpha = 2^o$, acquisition delay (i.e. echo time) = 300 μs, and TR = 2 ms. For a 64 x 128 x128 k-space matrix the total imaging time amounted to 35 min. At 3T, the voxel size could be reduced to 1.6x1.6x3.2 mm^3, and the acquisition delay could be shortened to 280 μs, because a 40 mT/m gradient strength was available. For reconstruction, data were directly copied into a Cartesian grid, and 3D Fourier transformation was applied.

3.1.2.3 Pointwise Encoding Time Reduction with Radial Acquisition (PETRA)

The imaging parameters for PETRA were set to the same values as UTE, except that the scan time was 50 s longer due to the single point data collection at the end, which corresponds to 5000 data points. As with UTE, image data were reconstructed by gridding with a subsequent Fast Fourier Transform.

3.1.3 Hardware Optimization for MRI of Ultra-short T_2 Samples

Ultimate aim of the resonator optimization is to have a high sensitivity antenna with a homogeneous magnetic field distribution. Additionally, ringdown time of the resonator and Tx/Rx switching times should lie within acceptable range not to increase the acquisition delay. To this end, solenoid, loop-gap, and birdcage resonators were built, and their homogeneity and ring-down times were compared.

3.1.3.1 RF Resonators

3.1.3.1.1 Solenoid Resonators

A solenoid resonator design was not only chosen to maximize the filling factor, but also to achieve a high B_1 homogeneity over the resonator cross-section, which is critical for the T_1 relaxation time mapping. Solenoid resonators were built for two different clinical MR systems: 1.5 T Magnetom Tim Symphony, and 3T Magnetom Tim Trio, Siemens, Erlangen. Solenoid resonators were wound on an acrylic glass cylinder (inner diameter: 80 mm, wall thickness: 10 mm, length: 180 mm, 7 turns, cf. Fig. 35a) with copper adhesive tape (3M 1182, 13 mm width). The resonators were tuned and matched to the resonance frequencies at 1.5 T and 3 T (i.e., 63.71 MHz and 123.29 MHz) yielding unloaded *Q* factors of 55 and 158, respectively inside the MR scanner (The existing conducting parts inside the magnet bore degraded the *Q* value). Note that, loading the resonators with the dehydrated mummified sample did not change *Q*. The ring-down time was measured using a pick-up coil and an oscilloscope (Tektronix MDO3104, Oregon, USA) as 3.4 μs / 3.2 μs for 1.5 T / 3 T designs, respectively. Design of the resonators, S11 plots, and the corresponding FDTD simulations performed with Sim4Life (Zurich Med Tech, Zurich, Switzerland), where B_1 field maps in axial plane are shown in Fig. 35a,d,g.

To quantify the resonator homogeneity, the resonators were loaded with homogenous distilled water phantoms, and $|B_1|$ was calculated using the double-angle method [76].

3.1.3.1.2 Loop-gap Resonator

A loop gap resonator at 3 T resonance frequency was constructed using a copper foil with adhesive tape (3M 1182, 160 mm width) wound around a cylindrical acrylic glass of the same physical dimensions as in solenoid design (Fig. 35b). Unloaded *Q* was calculated as 160 from the S11 measurements. The ring-down time was measured as 2.6 μs. A photo of the resonator, S11 plots, and the corresponding B_1 field map simulations in axial plane are shown in Fig. 35b,e,h.

Loop-gap resonator consists of a large area of conductor that makes it susceptible to eddy current induction effects due to switching gradient fields. To reduce the effect of eddy currents, the copper foil was galvanically separated into 8 smaller parts using DC blocking capacitors with high capacitance (5.6 nF).

3.1.3.1.3 Birdcage Resonator

A low pass quadrature birdcage resonator with 8 rungs was designed for 3T resonance frequency. The resonator was constructed on an acrylic cylinder with the same dimensions as the solenoid resonator using adhesive copper tapes (3M 1182, 13 mm width) for the rungs and the end rings. Length of each rung was 90 mm. An RF shield of diameter 200 mm was also equipped in design. Unloaded Q was calculated as 112. The ring-down time was measured as 3.1μs, A photo of the resonator, S11 plots, and the corresponding B_1 field map simulations in axial plane are shown in Fig. 35c,f,i.

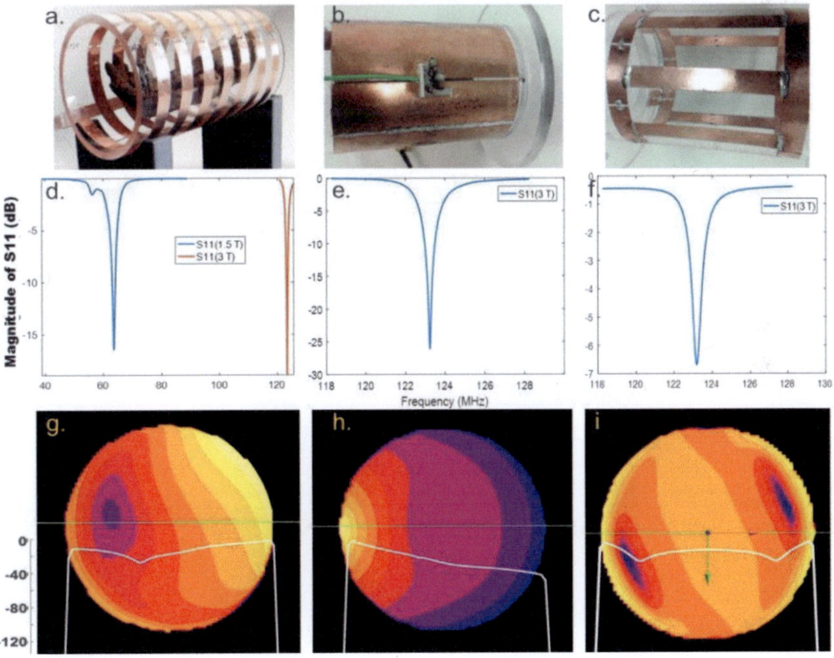

Fig. 35: Photos, reflection characteristics, and electromagnetic field simulations for three RF resonators. Photos of solenoid (a.), loop-gap (b.), and birdcage resonators (c.); corresponding S11 magnitude plots (d.),(e.),(f.); and simulated B_1 field maps in axial plane with the line profile overlaid (g.),(h.), and (i.), respectively. Note the different axis ranges in d-f.

During MR imaging, the resonators were placed at the system's iso-center using a PVC holder to avoid off-center shifts in the subsequent radial image acquisitions [184].

3.1.3.2 RF Circuitry: Tx/Rx Switches

To connect the resonators to the MR systems, dedicated Tx/Rx switches were constructed (Fig. 36). The Tx/Rx switches consisted of two lumped element $\lambda/4$ networks (π configuration), and they were operated with PIN diodes (MA4P7446F-1091, M/A-COM Technology Solutions, Inc., Lowell, USA). The Tx/Rx switches provided up to 50 dB isolation, which was achieved by placing two PIN diodes in series.

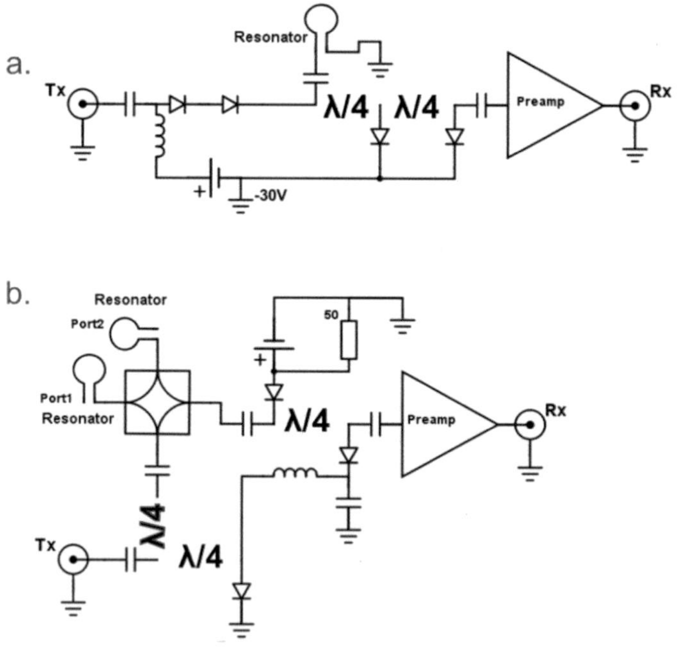

Fig. 36: Tx/Rx switch with high isolation and rapid switching time for linearly polarized single channel resonators (a.), and circularly polarized quadrature driven resonators, which uses a hybrid coupler to drive two ports of the resonator with $\pi/2$ phase difference (b.)

The PIN diodes were actively switched by the resonator control signals obtained from the MRI system. As all imaging sequences aim at using very short echo times, special care was taken to achieve short switching and ring-down times. The PIN diodes were selected to have a sufficiently low carrier life time of 19 μs, to sustain high power of 25 W, and low resistance of 0.5 Ω. Additional measures to enhance switching performance were using cross diodes to increase isolation, and additional $\lambda/4$ lines to decrease switching time.

For the details of the circuit schematic, PCB layouts, and test bench measurements for PIN diode Tx/Rx switches for linearly polarized single channel resonators and for quadrature resonators, please refer to Appendix D.

3.1.4 Parametric Mapping and Data Analysis

The mummified hand was imaged using the solenoid resonators with all three sequences at both 1.5 T and 3 T with almost identical protocol setups. Sequences were compared in terms of SNR. T_1 and T_2* mappings were generated using UTE sequence. For anatomical reference, all image data were compared to a CT image set of the specimen.

3.1.4.1 SNR Analysis

To compare the sequence performance, SNR maps were calculated by dividing the images by the standard deviation calculated in a noise-only region of the image. For various anatomical structures, SNR was calculated as the average signal in manually-selected regions of interest (ROI) covering typically four pixels or more. Differences in the imaging parameters were accounted for in the SNR calculations which describe SNR of a single pixel in the image as in Eq.2.3.1.

Relaxation parameters were also analysed in different anatomical regions using the same regions of interest.

3.1.4.2 T_2* Relaxation Time Mapping

To determine T_2^*, a series of 8 3D UTE data sets were acquired with TE ranging from 50 μs to 0.9 ms (TR = 3.8 ms, base resolution = 176, FOV = 176x176 mm^2, radial spokes =40000, flip angle = 10°, and acquisition bandwidth = 859 Hz/pixel). T_2* values were calculated with a Matlab algorithm by fitting the square of the signal intensity values to a mono-exponential decay function with a time constant of $2/T_2$*, plus an arbitrary constant representing the noise [185].

3.1.4.3 B_1 Mapping

Using the double-angle method described in section 2.2.1.6, flip angle values were calculated using two GRE images acquired with 5° and 10° flip angles for each resonator. The resonators were designed for unloaded condition considering the highly dehydrated load. Therefore, 3 plastic tubes filled with distilled water were used as phantom in B_1 measurements to reduce the shift in tuning frequency and matching. (Matlab script in Appendix E). The reason for using plastic water tubes instead of a cylindrical phantom filling the whole volume is to minimize the conductivity and permittivity effects of the loading on the resonator properties, which are tuned and matched for the unloaded condition.

3.1.4.4 T_1 Relaxation Time Mapping

For T_1 measurements, UTE data were acquired with a series of 14 flip angles between 5° and 70°, and T_1 values were calculated using the FLASH angle equation explained in Theory, section 2.3.4.3. In this equation, transverse magnetization decay during RF excitation was taken into account. The sequence parameters were the same as for the T_2* measurements with TE fixed to 70 μs. Relaxation maps were generated for both magnetic field strengths, and an SNR threshold mask was applied for better visualization at

SNR > 5. Prior to the relaxation curve fitting, noise correction was applied as described in [186]. Another threshold was applied after fitting; all pixels with a root-mean-square error above 10% were discarded.

3.1.4.5 Sharpness Comparison

To quantify the sharpness (i.e., the capability for accurate edge detection) a rectangular prism-shaped rubber phantom with T_2* of 600 μs was imaged with UTE and SPI using the same imaging parameters as in the SNR measurements. Line plots of the sagittal cross-sections were compared with respect to their deviation from a 1D rectangular function.

3.2 Concurrent Excitation and Acquisition

3.2.1 Active Decoupling of Transmit and Receive Resonators using a Transmit Array System

In this section, the methods to implement active decoupling will be detailed. This work was partially used in a US patent application (US20130271135), and described in [142].

3.2.1.1 Decoupling of RF Resonators with Additional Decoupling Resonators

The resonator setup for the CEA experiments consisted of a primary Tx resonator (Tx coil-1), a decoupling Tx resonator (Tx coil-2), and a Rx resonator, which is the most simple implementation of proposed CEA concept (Fig. 37).

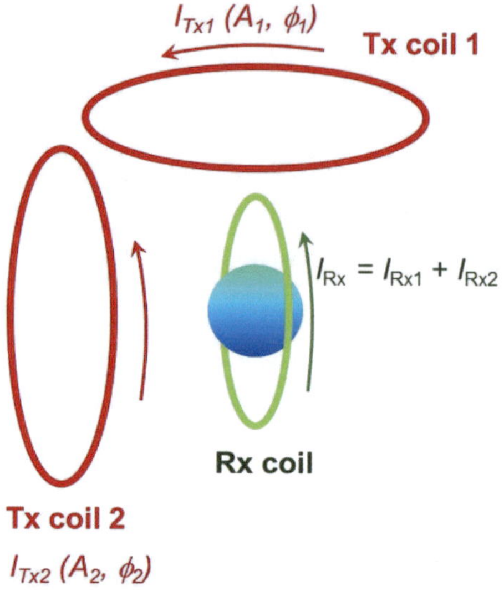

Fig. 37: Visualization of the decoupling process: Tx coil-1 (Tx resonator) and the Rx resonator are placed orthogonally, which reduces the B_1-induced current on the Rx coil. Tx coil-1 is driven by a current I_{Tx1}, which induces I_{Rx1} on the Rx resonator. The amplitude and phase of Tx coil-2 (decoupling resonator) are adjusted such that I_{Rx2} cancels I_{Rx1} out, which significantly reduces the total B_1-induced voltage. Because I_{Tx1} is greater than I_{Tx2} by the amount of geometrical decoupling, the spins of the sample are excited mostly due to the RF magnetic field produced by Tx coil-1.

All RF resonators were constructed from adhesive copper tape (3M 1182, 8 mm width) on proton-free glass plates to eliminate the background signals from polymers with very short T_2* values (Fig. 38). The octagon-shaped RF resonators have diagonal lengths of 10 cm and 15 cm for the Rx resonator and the Tx resonators, respectively. For 3D MRI experiment, circular RF resonators which have diameters of 5 cm and 10 cm for the Rx resonator and the Tx resonators, respectively, were used. All resonators were initially tuned to the proton resonance frequency at 3 T of 123.3 MHz.

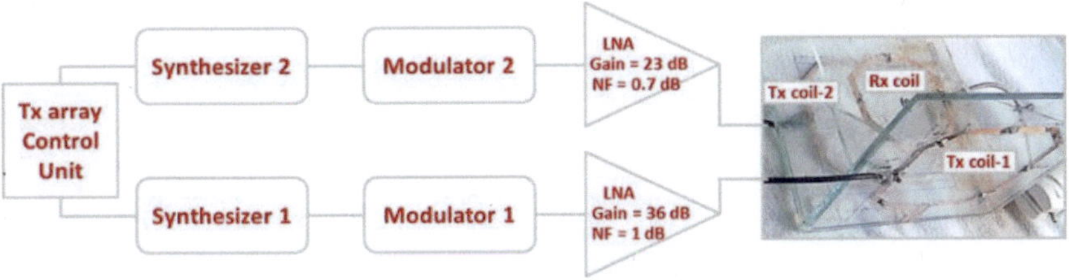

Fig. 38: Schematic of the system setup. The Transmit Array control unit controls the amplitude and phase as well as the envelopes of the RF waveforms via a modulator. The output of the one of the modulators is amplified by a homemade LNA. An oscilloscope is used to check the voltage levels at the modulator outputs and at the Rx coil output in real-time. Once the signal voltage level at the Rx coil output is reduced to the dynamic range of the Rx circuitry of the scanner, the Rx coil was connected to the scanner via a homemade ultra-low noise preamplifier (ULNA).

Several measures were taken to maximize decoupling between the Tx resonators, and between Tx coil-1 and the Rx coil:

1. The Tx resonators 1 and 2 were decoupled from each other by capacitive decoupling [187]. Therefore, the resonators were connected (see connection between Tx coil-1 and 2 in Fig. 38) and a capacitance was added so that the coupling parameter S12 was minimized. With a value C = 88 pF, a decoupling of 15 dB was achieved.
2. Tx coil-1 and the Rx coil were decoupled geometrically by placing the coil planes orthogonal to each other, with the Rx coil at the mid-plane of Tx coil-1 (cf. Fig. 37), which in the ideal case nulls the magnetic flux in the Rx coil. In practice, the field polarizations are not ideally linear [72] and the fields are distorted by the presence of the imaged object, which limit the achievable amount of decoupling.
3. The remaining B_1-induced current of Tx coil-1 in the Rx coil is cancelled by adjusting the phase and amplitude of the transmit field from Tx coil-2 (PA decoupling). Note that the geometrical decoupling between Tx coil-1 and the Rx coil has already significantly reduced the B_1-induced transmit signal, and thus Tx coil-2 can be driven at an input power level that is much lower than that of Tx coil-1.

4. In addition, the decoupling resonator Tx coil-2 was slightly detuned (Δf = 2 MHz, S11(123 MHz) = -3.2 dB) to reduce mutual coupling with the Rx coil. If Tx coil-2 and the Rx coil are weakly coupled, the sensitivity of the Rx coil is maintained, as no resonance shift does occur.

Due to the low input power applied to Tx coil-2, the flip angle in the sample is mainly determined by the Tx coil-1, and the RF transmit field of Tx coil-1 is nearly unchanged. Thus, the magnetization in the sample is still excited by the transmit RF field and an MRI signal can be detected by the Rx coil even during RF excitation. In addition, the orthogonal placement not only reduces the RF signal coupling, but also noise coupling between the RF transmitter and the Rx coil.

Because the commercial RFPA of the Tx coil-1 induces high levels of noise up to 400 mV_{pp}, it was replaced by a homemade 100 mW low noise amplifier (LNA) with a gain of 15 dB and a noise figure of 3 dB. The LNA was designed as a single stage amplifier using a BFG135 bipolar transistor with input and output matching [188]. Please refer to Appendix F for detailed circuit design and test bench measurements for a bipolat transistor based LNA. Tx coil-2 required much lower RF power levels, and was thus directly connected to the modulator output of the second transmit channel (Fig. 38).

The following steps were performed to set up and characterize the resonator decoupling:

- To maximize the geometric decoupling, the total unloaded resonator setup was placed on the patient table at the iso-center of the magnet. Then the Tx coil-1 was connected to the RF transmitter via the LNA, and the RX coil was connected to a digital oscilloscope (DSO6104A Agilent). The Rx coil was then slowly rotated, and the amount of de-coupling was noted as a function of the relative angle between the resonators. The rotation was done first in steps of 5°, which was then refined to steps of 1° for fine tuning. Then the Rx coil was fixed at the angular position, in which the B_1-induced voltage was minimal.

- The phantom (see below) was then placed at the imaging volume of the setup for the subsequent PA decoupling calibration to take into account the effect of the conductive sample in the decoupling.
- To adjust the PA decoupling parameters c and θ in Eq. 2.4.2, RF power was individually applied to Tx coil-1, and to Tx coil-2. Then *c* was calculated from the ratio of the peak voltages, and the phase θ was first adjusted experimentally via the user interface of the parallel transmit system until the output voltage at the Rx coil was at the noise level of the oscilloscope output (V_{pp} = 2 mV).
- To increase the sensitivity, the Rx coil was then connected to the MR receiver via a homemade ultra-low noise preamplifier (ULNA, gain: 25 dB, noise figure: 1 dB). The decoupling was further improved by fine-adjusting the PA decoupling parameters c (sensitivity: 0.001/3 ± 0.5 dB for 4 V peak RF signal voltage) and θ (sensitivity: 0.01° / 3 ± 0.5 dB) iteratively at the Tx array console, until the received signal was close to the noise level. The degree of decoupling was plotted as a function of θ using both the oscilloscope data and the ULNA measurements.

After optimization of the individual decoupling steps, the decoupling schemes were combined to show that CEA is possible in clinical MR systems with minimal modifications.

Flip Angle

To provide absolute values for the flip angle, Klose's method was used [189], in which the magnetization was prepared with the chirp pulse, and the remaining longitudinal magnetization was read out as an FID (S_1). The experiment was repeated without preparation (S_2), and the effective flip angle was calculated from the signal ratio: $\alpha = \cos^{-1}(S_1/S_2)$.

Transmit Noise

To characterize the noise from the RF transmitter during CEA, RF noise was measured as the standard deviation of the acquired CEA signal from a rectangular RF pulse. Data were acquired for different decoupling conditions: PA decoupling with 20 dB, 10 dB and no geometric decoupling. Signal noise was measured for an RF pulse with a U = 0 V amplitude and for a resonator setup with the transmit sources unplugged.

3.2.1.2 Application to MRI with CEA

All CEA experiments were conducted using a Siemens (Erlangen, Germany) Magnetom 3T clinical MRI system with an eight channel parallel transmit array unit. The parallel transmit system was used because it allows for an independent control of multiple Tx channels, and in order to facilitate RF supervision of multiple Tx channels the system allows for a concurrent signal reception during RF excitation, which is not possible with conventional MRI hardware of the same manufacturer.

The resonator setup was loaded with a cylindrical glass of tap water (glass diameter: 50 mm). After the decoupling procedure was carried out as described above, CEA projection data were acquired in the presence of a gradient (Fig. 39). Therefore, a chirp RF pulse of U_{peak} = 1 V was applied with a duration of 8.2 ms and a 8 kHz sweep range, and the gradient amplitude was set to 10 mT/m. Using the B_1 mapping data, this pulse provides a flip angle of 12° at the center frequency. The data acquisition was started 100 μs after the beginning of the chirp pulse, and 256 data points were sampled over a total period of 8 ms. To be able to subtract remaining background signal from insufficient decoupling, the experiment was repeated without the sample to acquire the system response *h*(*t*) (cf. section 2.4.3.).

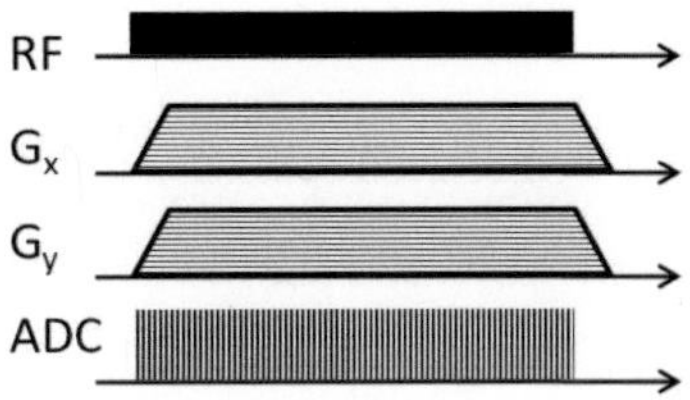

Fig. 39: 2D non-selective pulse sequence diagram for a single TR interval. The gradient strength and the sweep rate determine the FOV and resolution.

During subsequent imaging reconstruction, first, the reference raw data without sample were subtracted from the corresponding data taken with the sample. Second, a deconvolution was performed as outlined in Eq. 2.4.7 resulting in radial projection raw data, as it would have been acquired with a conventional 2D radial MRI sequence. The deconvolved data were then mapped onto a Cartesian grid using re-gridding, and an inverse Fourier transformation was applied [190]. Data processing from raw data to the FID and the projection data is presented for a water sample.

In a second experiment the resonator setup was loaded with a thin rubber sample (5x3x0.8 cm^3), which was chosen because it has a very short T_2* of 500 μs (measured independently). Several holes (diameters: 0.1-0.4 cm) were drilled into the sample to provide an inner structure. The acquisition was repeated by rotating the gradient direction radially over 768 angular values (2D radial sampling), resulting in a total acquisition time of 768 x 200ms = 2:34 minutes. An image of the rubber phantom was reconstructed from the CEA data as described above.

CEA pulse sequence was modified to employ 3D radial k-space trajectory [56] (cf. section 2.1.6). *Ex vivo* adipose APOE mouse (65 g Body weight) was imaged to demonstrate that the setup can also be used with more realistic objects. The setup was improved with HD24388 (gain: 23 dB, noise figure: 0.7 dB) and HD29980 (gain: 36 dB, noise figure: 1.0 dB) amplifiers (HD Communications Corp. Ronkonkoma, NY) at both transmit channels. 100000 radial spokes were acquired during chirp RF excitation spanning 40 kHz over 4 ms. In time interleaved mode, this pulse produced 9° flip angle at a homogenous phantom. Maximum gradient strength was set to 24 mT/m, and 512 points per spoke were acquired. Spoiler gradients of 24 mT/m amplitude and 4 ms duration were applied following each acquisition. TR of 20 ms resulted in total acquisition time of 33 min. The image is reconstructed onto 512 x 512 x512 Cartesian grid using a Gaussian gridding kernel. For anatomical reference a 3D GRE sequence with FOV = 120 mm, TR = 7 ms, TE = 2.21 ms, flip angle = 30°, 400 Hz/Px bandwidth, 0.4 mm resolution and 0.5 mm slice thickness is applied.

The effect of the transmit system instability on the acquired signal after decoupling is also demonstrated by a CEA experiment. The B_1-induced leakage signal is acquired at 1 min steps while keeping the entire sequence parameters constant using a CEA setup with no sample present.

3.2.2 Automated Analog Cancellation using Dynamic Feedback from MR

CEA with active analog cancellation and real-time feedback was implemented in a clinical 3T MRI system (Siemens, Erlangen). The setup of the total system is shown in Fig. 40: The transmit (Tx) power is delivered to the main Tx resonator via a directional coupler, where a small fraction of Tx signal is fed into a phase shifter and amplitude attenuator. This phase-shifted and attenuated copy of the Tx signal is then subtracted from the receive signal to cancel the unwanted leakage of Tx signal into the Rx resonator. The settings of the phase shifter and the attenuator are continuously updated in a real-time feedback loop.

To keep Tx noise low, a custom made LNA with 30 mW output power was used in Tx chain. The Tx loop resonator (ø = 15 cm) and the Rx loop resonator (ø = 12 cm) were placed orthogonal to each other for geometric decoupling. In addition, a small pick-up loop coil (PUC) was placed in close proximity to the Tx resonator to independently measure the Tx waveform.

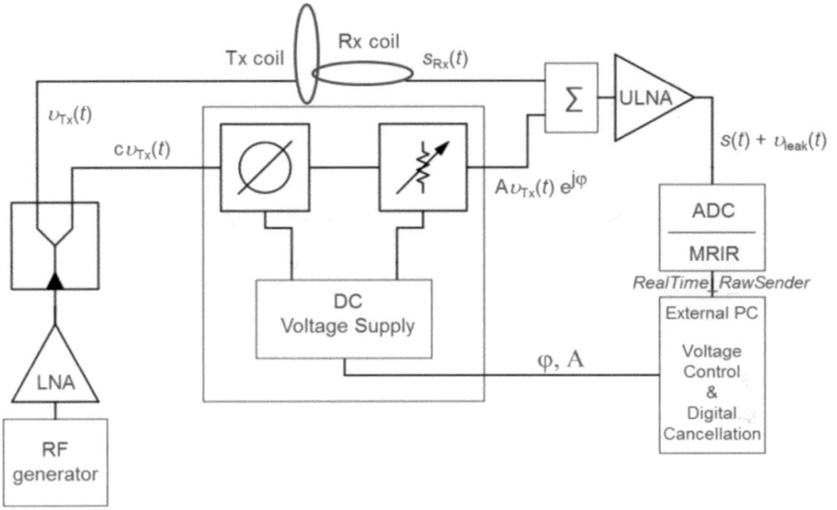

Fig. 40: System setup for analog cancellation with real-time feedback. An external PC controls the phase and attenuation level of a small copy of the Tx signal to minimize the mean of the residual leakage signal. This control is based on serial communication via USB connection. Data transfer from MR image reconstruction unit (MRIR) to the external PC is realized by TCP/IP protocol.

3.2.2.1 Analog Cancellation Circuit

The analog cancellation concept is based on subtracting a small copy of the Tx signal from the measured Rx signal at the Rx resonator output after proper phase and amplitude adjustment. A directional coupler (ZDC-10-1+ Mini-Circuits, Brooklyn, NY) divides the signal into two parts: the Tx voltage signal, $\upsilon_{Tx}(t)$, and the correction signal, $c\upsilon_{Tx}(t)$ where c = 0.28. During continuous RF excitation, a certain amount of Tx power is coupled into the Rx resonator even if geometrical decoupling is used to minimize resonator cross-talk. The task of the analog cancellation system is to match the phase φ_{cpl} and the amplitude A_{cpl} of the Tx-induced signal at the Rx resonator by searching for the optimal phase shift and the attenuation (Eqn.1).

Then, the analog signals are subtracted using a power combiner (ADP-2-1+ Mini-Circuits, Brooklyn, NY) before they are amplified and digitized in the subsequent Rx chain.

$$s_{Rx}(t) = A_{cpl} u_{Tx} e^{j\varphi_{cpl}} + s(t) \tag{3.2.1}$$

$$s_{Rx}(t) - A u_{Tx} e^{j\varphi} = s(t) + \min_{A,\varphi} u_{leak} \tag{3.2.2}$$

3.2.2.1.1 Voltage controlled phase shifter

Analog phase shifting was realized by a reflection type phase shifter [191] where impedances at 2nd and 3rd ports of a 90° hybrid coupler (HE128MF EMC Technology, Florida, USA) were changed using varactor diodes (BB640 Infenion Technologies, Munich, Germany). A digital programmable power supply (HMP4040 Rohde&Schwarz, Munich, Germany) was used to control voltage across, and thus the capacitance C_{var} of, the diodes – according to the manufacturer data and own measurements, up to a voltage of 25 V the varactors' capacitance monotonically increases to 75 pF. A simplified circuit schematic and the resulting phase shift are shown in Fig. 41.

3.2.2.1.2 Voltage Controlled Attenuator

The variable attenuator was designed based on a π-network formed by three PIN diodes (MA4P7446F-1091, M/A-COM Technology Solutions, Inc., Lowell, USA) which were controlled by a fixed and a variable voltage source. PIN diodes behave as voltage controlled resistors. A simplified circuit schematic is shown in Fig. 42. As V_{att} increases, the resistance in RF path decreases, and so does the attenuation. DC blocking capacitors are to prevent DC signal leak towards other parts of the circuit and they have a 2.7 nF capacitance. The inductor together with two shunt capacitors are included to prevent signal leakage from D_1 to D_3 that would impair the performance of the circuit.

The maximum attenuation is achieved when V_{att} is at a minimum. At this point, current from the supply V_{fix} turns the diodes D_1 and D_3 on effectively shorting the signal to ground. D_2 is then reverse biased. When V_{att} is above the turn on voltage plus V_{fix}, D_1 and D_3 become reverse biased, and D_2 becomes forward biased, allowing the signal to pass through the circuit. The fixed voltage was set to V_{fix} = 4.5 V. At V_{att} = 8.0 V, a minimum value of 1.3 dB attenuation was obtained (cf. Fig. 42). Between V_{att} = 2.5 V and 8.0 V the attenuation increases nearly linearly at about 3.0 ± 0.3 dB/V.

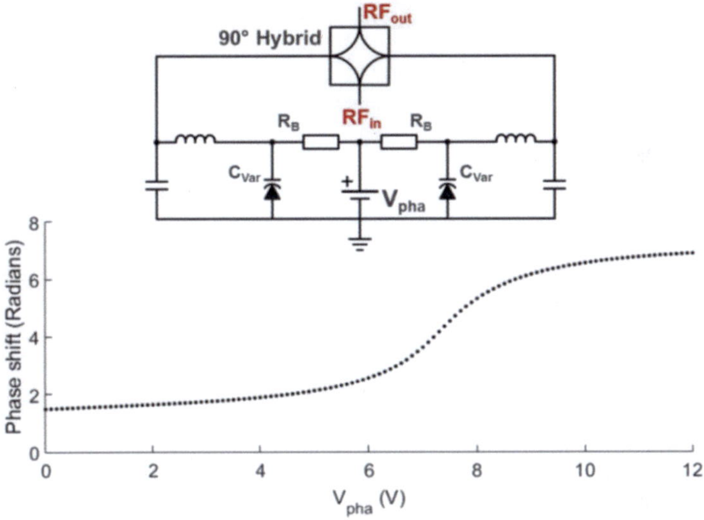

Fig. 41: Voltage controlled phase shifter simplified circuit and voltage vs. phase plot. R_B stands for bias resistors. Phase shifter enables 280° coverage for 0 V < V_{pha} < 12 V. Nonlinear behavior is taken into account by using variable voltage step sizes in automated feedback decoupling.

In Fig. 43, measurements with network analyzer (ZVB4, Rohde&Schwarz, Munich, Germany) show that the combination of analog cancellation system and geometrical decoupling yields 60 to 72 dB decoupling over 60 kHz band around the center frequency, of which, 48 dB is provided by the analog cancellation system alone. Input and output reflection coefficients, |S11| and |S22|, are larger than 12.5 dB over a frequency range of 200 kHz, which is larger than the acquisition bandwidth used later for imaging of about 16 kHz.

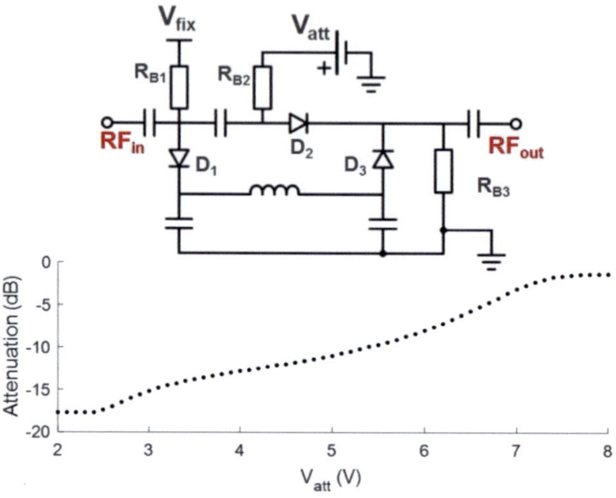

Fig. 42: Voltage controlled attenuator circuit schematic and voltage vs attenuation in logarithmic scale. $R_{B1,2,3}$ are the bias resistors. Attenuation/Voltage ratio is 3.0 ± 0.3 dB/V for V_{att} = 2.5- 8.0 V, where V_{fix} = 4.5 V. Bypass capacitors, RF chokes and the impedance matching elements are excluded for simplicity.

Please refer to Appendix G for circuit diagram, PCB layout, and simulation results of the analog cancellation circuit.

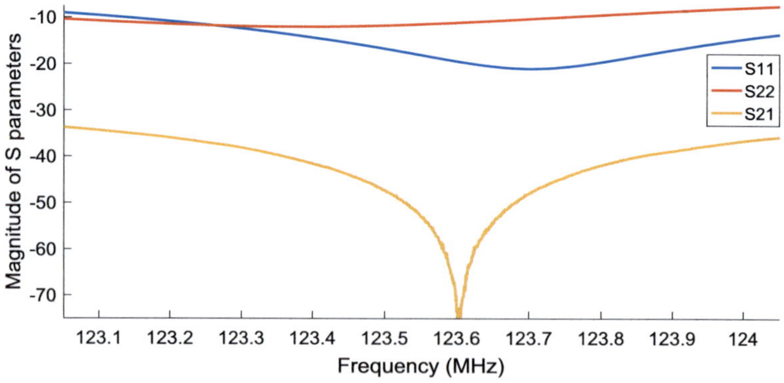

Fig. 43: Network analyzer measurement results for manually tuned analog cancellation circuit using the experimental resonator setup.

3.2.2.2 Real-time feedback operation

An external PC was used to implement real-time feedback control of the phase shifter and attenuator units. Raw data was acquired with a special calibration pulse sequence that consists of a constant rectangular RF pulse of 4 ms duration, a concurrent data acquisition, and a subsequent spoiler gradient. MR raw data from each repetition was sent directly from the MRI receiver system to an external PC using real-time data transfer routines [192] within the MR system image calculation environment (ICE). On the PC a control program written in MATLAB was then changing the phase and amplitude settings of the analog cancellation circuit via a USB interface which provided control voltages for the power supply of the analog cancellation circuit. The calibration sequence was repeated until the residual transmit leakage was below a user-defined threshold – typically, a threshold value of 80 dB was used.

During automatic optimization a gradient descent approach was used to change phase and amplitude values. Initial voltage levels were determined experimentally to avoid receiver saturation, and then at first the phase was changed using a voltage step size about 0.3 V (cf. Fig. 41), and the step size was lowered to 0.1 V until a minimum was reached. Next, the attenuator was iteratively adjusted using an initial step size of 0.05 V (cf. Fig. 42). In subsequent iterations step sizes down to 0.005 V were applied until the acquired signal intensity was below the specified threshold. In this preliminary implementation the total time for one feedback cycle was 50 ms.

In the 3T MRI system, dynamic decoupling performance was tested under real-time feedback operation. In the experiment with the hand, initial decoupling took 40 iterations or 2 s to converge, and in other experiments at maximum 100 iterations were needed. Subsequently, fewer iterations were required to re-adjust the parameters after the position of the hand was changed intentionally (Fig. 44, arrows). The performance was not affected when the threshold was lowered from 78 dB to 89 dB on-resonant isolation.

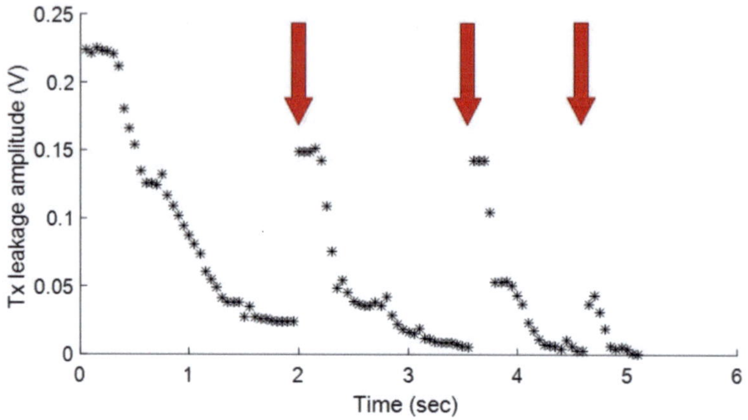

Fig. 44: Voltage level of the residual Tx induced leakage signal during automated feedback cancellation. Arrows show the instants when the hand position was changed and re-iteration was triggered.

To demonstrate the performance of the analog cancellation system, the resonators were loaded with a human hand, and the position was changed three times during the calibration procedure and the signal intensity levels were recorded.

To demonstrate the temporal stability of the system during MR imaging, the coupling of the unloaded resonator system was first minimized with the calibration sequence, and then a concurrent excitation and acquisition sequence with 100 radial spokes was acquired with TR = 6 s using a smooth-chirp excitation with 16 kHz sweep range. Measurements were repeated with zero RF voltage to analyze the noise stability – therefore, the standard deviation of the acquired signal was calculated for each repetition. As the reference receive noise floor, an acquisition was performed when the Tx cables were unplugged.

An additional performance test was conducted to measure the acoustic noise performance of the CEA pulse sequence. A Digital sound level meter (DSL 331, Tecpel Co., Ltd.,Taiwan) was placed at a 2 m distance to the magnet, with the calibrated microphone pointing towards the magnet bore. Background sound level, CEA sequence with 60,000 spokes, UTE sequence with the same number of spokes for G_{max} = 20 mT/m, and for G_{max} = 10 mT/m, and GRE sequence with the reported parameters in *in vivo* measurement (cf. section 3.2.2.4.) were compared in terms of acoustic pressure levels in dBA.

3.2.2.3 Alternative resonator setup

In the alternative CEA setup the pin-diode variable attenuator was replaced with a matched hybrid coupler attenuator (Fig. 45a), so that the analog cancellation circuit has also a broader bandwidth in terms of matching. PIN diodes were connected in parallel to a dummy load, because this configuration already provides sufficient attenuation and it is less sensitive to voltage variations within the required attenuation range (cf. section 2.4.2.1.2.). The voltage vs. attenuation plot for fixed phase at V_{pha} = 3 V are shown in Fig. 45b.

Phase shifter was also modified to have a more linear and less voltage-sensitive response within the required phase range. The bias resistance was increased and the capacitance range was adjusted by adding another varactor diode in parallel to the existing ones (cf. Fig. 41). New phase response at fixed attenuation is shown in Fig. 45c.

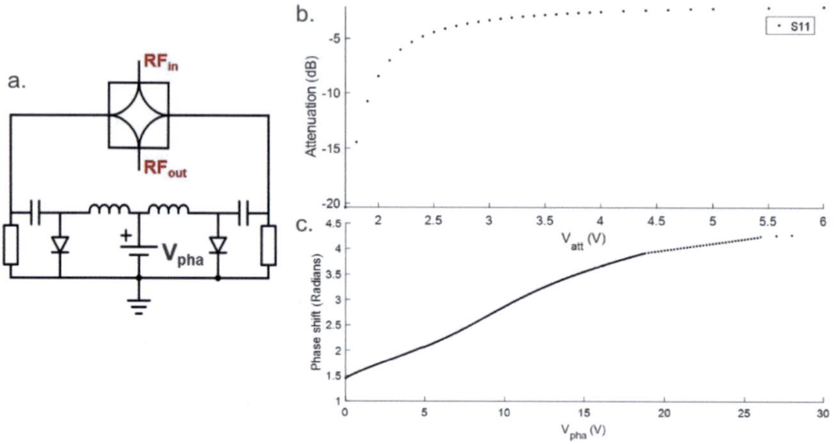

Fig. 45: Hybrid coupler based voltage controlled reflection type attenuator circuit design (a.), and the voltage vs. attenuation response for fixed phase shift (b.). Voltage response of the modified phase shifter (c.)

Since the new analog cancellation circuit has two intrinsically matched parts, the reflections are minimized at the input and output ports ($S11_{max}$ = -19.5 dB, $S22_{max}$ = -21.7 dB). And due to decreased voltage sensitivity of phase and attenuation settings, higher isolation levels can be obtained. Isolation test using the new analog cancellation system with the geometrically decoupled resonator pair is shown in Fig. 46. The matched analog cancellation setup reduces the number of voltage sources to two, by eliminating the fixed voltage source used in the pin-diode based attenuator (section 3.2.2.1.2.).

Circuit diagram and simulation setup of the matched analog cancellation circuit is given in Appendix G.

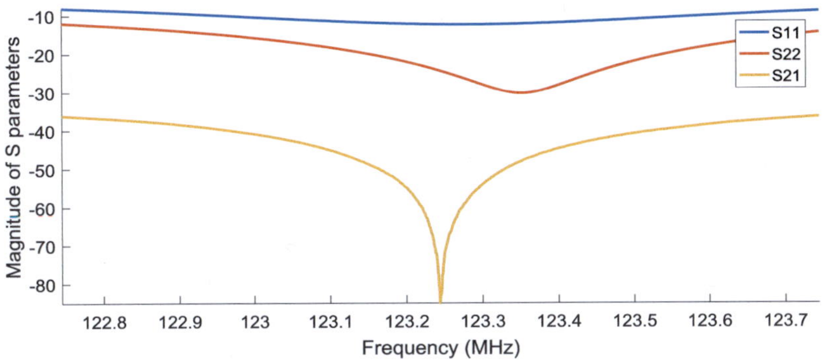

Fig. 46: Network analyzer measurement results for the analog cancellation circuit with matched attenuator using geometrically decoupled resonator setup. Isolation is higher (i.e. |S21| is lower) with the new analog cancellation circuit.

3.2.2.4 Application to MRI with CEA

To test the system, an imaging experiment was performed with a phantom consisting of 8 cylindrical tubes filled with a solution of 0.5 g/l $NiSO_4$. Additionally two smaller tubes filled with 1 g/l and 2 g/l $CuSO_4$ solutions were attached using plasticine based industrial dough. A 3D radial CEA imaging sequence (Fig. 47) with continuous gradient-ramp [143] was applied with the following parameters: 59999 radial spokes, 512 data points per spoke, TR = 8.2 ms, acquisition window = 4.1 ms, maximum gradient strength = 4 mT/m, chirp RF pulse with 16 kHz sweep. The chirp pulse amplitude was smoothed by a sinusoidal function at the first and last 10% of the complete pulse duration [46] to avoid distortions in acquired signal due to sudden onset of the RF signal. The peak power at the output of the power amplifier was measured as 17.3 mW.

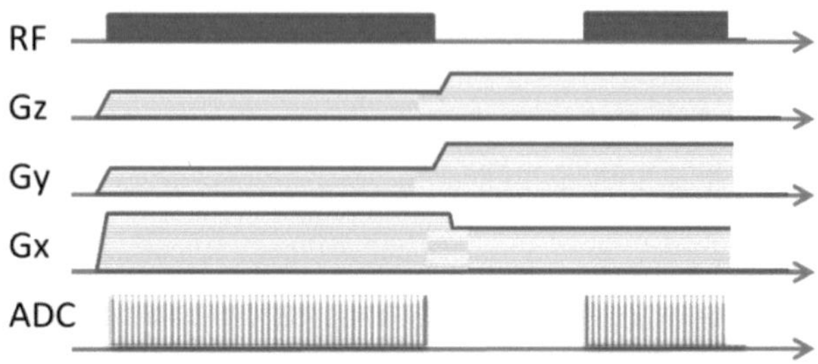

Fig. 47: Improved 3D radial CEA pulse sequence. Continuous gradient-ramps minimizes the acoustic noise. ADC and RF pulse events start after the gradient-ramp: gradient delay based systematic errors and ramp-up sampling are avoided.

For each radial spoke, residual leakage was subtracted digitally and deconvolution with the input chirp signal was performed to obtain projection data as described in section 2.4.3. A k-space center shift correction was applied based on the cross correlation of the opposite radial spokes to account for sub-pixel shifts in the calculated FID similar to the gradient delay calculation used in conventional radial imaging [182]. Finally, 3D non-uniformly sampled FID dataset was mapped on a Cartesian grid using Kaiser-Bessel interpolation [190] and Fourier transformed to obtain image data. All calculations were implemented in MATLAB 9.0 (The MathWorks, Inc., Natick, Massachusetts, United States). Matlab scripts are briefly described in Appendix H.

3.2.3 T_1-Modulation Effects of Frequency Swept MRI with CEA

CEA sequence with a radial inside-out acquisition scheme was simulated for 3 different k-space ordering methods: (i) Equidistant spokes based on the algorithm developed in [56], (ii) segmented equidistant trajectory with 16 segments, and (iii) the golden angle trajectory. A point source was defined on a 2D grid, and for each radial acquisition, the effective TR, TR_{eff}, was calculated as a function of distance from iso-center, d_n and the default TR (here: 4.2 ms), assuming a chirp excitation with a sweep range of f_0 - Δf to f_0 + Δf over 2 ms along with a constant gradient. Note that TR_{eff} changes from one radial spoke to the next for all off-center positions. Transverse magnetization was prepared to reach steady state after 100 dummy scans, then, modulations in transverse magnetization (M_{xy}) were calculated iteratively as a function of TR_{eff}, T_1 and the flip angle, α:

$$\mathrm{M}_n = \mathrm{M}_{\mathrm{n}-1}\cos(\alpha)\mathrm{e}^{-\mathrm{TR}_{\mathrm{eff}}(\mathrm{n}-1)/\mathrm{T}_1} + (1 - \mathrm{e}^{-\mathrm{TR}_{\mathrm{eff}}(\mathrm{n}-1)/\mathrm{T}_1}) \tag{3.2.3}$$

$$M_{xy,n} = M_n \sin\alpha \tag{3.2.4}$$

as was shown in [193].

Using the central slice theorem [194], 2D k-space is formed by weighting each radial spoke signal by the modulation in the magnitude of the transverse magnetization as a result of the variation in TR. Image reconstruction with zero-order gridding was used to obtain PSF of the point source located at the edge of FOV on a 512x512 grid. In extension to 3D calculations, similar steps were followed and PSFs were calculated. Segmentation and golden angle ordering was applied only along the azimuthal angle. Resulting k-space data was reconstructed using gridding with Kaiser-Bessel interpolation. PSF comparison was based on scaling and subtracting the reconstructed data obtained with T_1-modulated magnetization values and data reconstructed under constant TR assumption.

3.3 Electroglottography Measurements during Dynamic Lung MRI

With new clinical applications, new sensor devices for physiologic measurements are introduced into the MR environment. Physiological signals such as respiratory and cardiac data have been acquired since 1986 via pneumatic respiratory bellows and photoplethysmography (pulse oximetry) or electrocardiogram (ECG) leads [195]–[197]. Electrophysiological recording units were initially used to monitor the patient status, and, later, they were utilized in the imaging sequence for gating purposes [198]–[201]. In 1993, an MR-compatible electroencephalography (EEG) unit was marketed to study the electrical brain activity [202], [203], and in 1996 electrooculograms (EOG) were recorded during MRI to observe eye movements simultaneous with the functional MRI data [204].

In this work, another electrophysiological measurement device, an electroglottography (EGG) unit, is modified for use in the MR environment.

The purpose of this work is to combine information on the glottal cycle from EGG with anatomical information from MRI to study the physiological processes during singing. Therefore, a commercially available EGG device was modified to ensure functionality and safety during dynamic lung MRI. The performance of the device was evaluated in a phantom and a volunteer experiment, and a possible application in a combined lung MRI study with simultaneous EGG measurement is demonstrated.

3.3.1 EGG Device

For simultaneous MRI and EGG acquisitions, a commercial EGG device (EGG-D400, Laryngograph Ltd., London, UK) was modified to reduce RF interferences and to ensure safe operation during MR imaging. The control box of the EGG device was placed outside of the Faraday cage, and the two gold-plated copper electrodes consisting of an oval conductor and a circular ground plate (Fig. 48) were connected to the control box via extension cables. In the Faraday cage, the extension cables were modified by surrounding them with an additional copper mesh which was electrically connected to a waveguide of the Faraday cage. At the electrodes the shield was opened to enable electrical contact between the electrodes and the patient's skin. Through this opening, RF signal from the RF pulses during spin excitation can couple to the electrodes and contaminate the EGG signal. The shielding efficiency was tested by acquiring gradient echo (GRE) noise images without RF excitation at a high acquisition bandwidth of 130 kHz without the electrodes ('no device'), with the electrodes placed in the MR system but the device switched off ('device off'), and with an active EGG device ('device on').

To suppress RF-induced voltages in the EGG signal, a passive 4-stage Chebyshev RF filter (cut-off frequency = 8 MHz, stop band = 15 MHz with 70 dB suppression) was introduced into the receive path of the EGG device at the waveguide feed-through. RF-induced voltages were measured for different orientation of electrodes, different flip angles and were compared to the applied RF pulse shape which was monitored directly at the modulator output using an oscilloscope (Tektronix MDO3104, Oregon, USA). Note that the RF pulses add to the EGG signal as rectified envelopes (cf. Results).

To suppress possible shield currents, and, thus, to minimize RF heating of the shield, two floating BALUNs were constructed and placed on the shielded EGG cable [205]. Additionally, in order to keep transmit and receive channel grounds galvanically separated, 5 capacitors (1 nF) were introduced into the shield near the electrode tips.

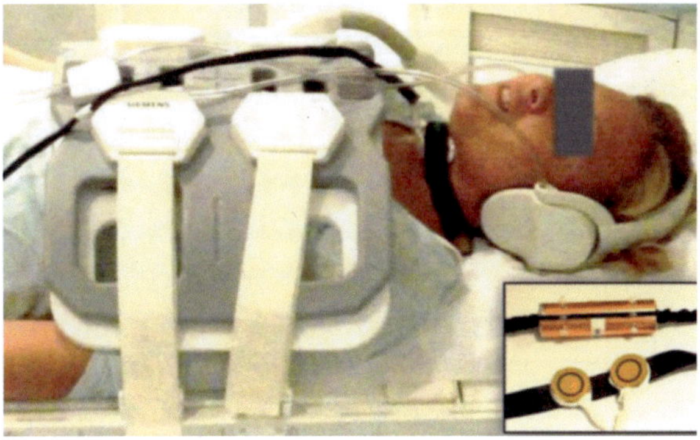

Fig. 48: The electrodes have 30 mm of total diameter and the optical temperature probes were attached to the tip. Electrodes are placed on two sides of the neck so that the electric currents cover the vocal folds. In practice the exact positioning is determined by palpation of the thyroid cartilage.

3.3.2 Signal Processing: Independent Component Analysis

Remaining RF interferences in the EGG signal were cancelled retrospect-tively. Therefore, an independent component analysis (ICA) [206]–[211] correction algorithm was implemented in Matlab® (2014b, The MathWorks, Inc., Natick, USA). At first, a band-pass filter (Low-high cut-off frequency = 20-2000 Hz, Bessel's filter of 9th order) was applied to the EGG signal, which was then separated into smaller time segments of 200±20 ms duration.

In conventional ICA, two statistically independent acquisitions (i.e., receive channels) are needed to separate two mixed signals, but in our setup the EGG device provides only single channel data. However, the EGG data are acquired under two different conditions: (a) during MR imaging alone, and (b) during MR image acquisition and with simultaneous vocal cord activity (singing, cf. below). Thus, the different EGG time segments either contained only the unwanted RF-induced signal artifact, or a combination of the EGG signal during singing and the RF artifacts.

To remove the artifacts from the contaminated EGG signals by ICA, at first 20 time-shifted copies were created (up to ±10 time points shift) from the pure artifact data. These copies were required for the ICA as the exact position of the artifacts in the EGG data was unknown. Assuming that the EGG signal and the RF induced voltage signal and its shifted versions are statistically independent, independent signal components were calculated using a fast-ICA algorithm [209]. The algorithm searches for a mixing matrix between the artifact signal and the contaminated EGG signal that minimizes the mutual information based on hyperbolic tangent (*tanh*) nonlinearity. ICA calculation was performed over the 20 shifted versions of the pure RF artifact signal and the original one along with the contaminated EGG signal, i.e. on a 22x3200 matrix for each time segment. The output with the lowest mutual information (i.e., the one with the best subtraction of the artifact data) was determined. Out of the 22 resultant independent components, the EGG signal without RF interference was selected manually. With the maximum number of iterations set to 1000, it took 0.12 s to calculate independent components for an input matrix size of 22x3200.

3.3.3 MR Safety Tests

To assess the safety of the system, magnetically-induced displacement forces were measured at a 1.5 T clinical MRI system (Tim Symphony, Siemens, Erlangen) using established ASTM procedures [212]. The electrodes were suspended from a string and the deflection angle was measured in the magnet's fringe field. Magnetically induced torque [213] was assessed by placing the electrodes on rotatable plate on the patient table, and measuring the alignment at different table locations and table movements. Image artifacts caused by the presence of the electrodes in imaging field of view (FOV) were measured with 2D GRE and spin echo (SE) pulse sequences (GRE: TR/TE = 500/15, FOV = 320 mm, FA = 30°, Acquisition matrix = 256x256, BW= 130 Hz/px, Slice thickness = 5 mm, In-plane phase encoding direction: anterior – posterior; SE: TR/TE = 500/20, FA = 90°, transverse slices 30° and 45° to the surface of the electrodes are displayed as in GRE, – cf. Fig. 49) [214]. During imaging, the electrodes were immersed in a 1 g/L copper sulphate solution in nylon net. In the images a pixel was classified as being in the artifact if the signal intensity differed by 30% or more compared to a control measurement [214].

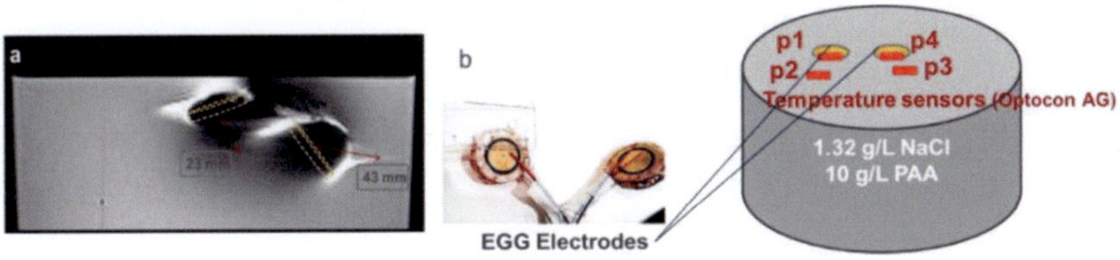

Fig. 49: The electrodes were dipped inside the container for artefact measuements. A slice from the SE image is shown with the position of the electrodes marked (a). The temperature probes attached to the surface of the electrodes and in close proximity (distance larger than 1 cm) of the electrodes (b).

RF-induced heating on or near the electrodes was tested for variable orientations of the EGG electrodes on a gel phantom [215]. Electrodes were supported with a plastic pipe holder and the locations with maximal heating were experimentally determined by testing for different alignment of the electrodes and searching with the temperature probes through roughly 5 mm step sizes around the electrodes. The electrodes were not completely immersed in the solution in order to create a similar positioning of the electrodes to that realized during EGG measurements (i.e., placement at the surface of the neck). Two optical temperature probes (Optocone AG) were attached to the electrodes, and another couple of probes were placed at variable positions in the vicinity of the electrode surface i.e. within 1 cm distance to the electrode surface. A high-SAR trueFISP sequence of 15 min duration and 2.1 W/kg SAR was applied (TR = 30 ms, FA = 66°, Time averaged RF energy [over 6 min) = 506 J/kg, Whole body averaged power = 17.8 W -vendor-reported values as taken from the user interface). Data from four temperature probes were recorded with 4.25 sec temporal resolution. The temperature was verified to be at steady state prior to the scan. After the RF energy was turned off, the temperature was monitored for at least two additional minutes. Room temperature was verified to be constant (20.4°C) during and within 15 min after completing the test.

3.3.4 Volunteer Measurements

After successful completion of the safety tests, simultaneous 2D trueFISP MRI with 3 frame per second (fps) (TR/TE = 3/1.5 ms; α = 6°; BW = 977 Hz/px; Slice thickness = 10 mm; Acquisition matrix = 256; FOV = 420 mm; SAR = 0.042 W/kg; Total acquisition time = 42 s; Time averaged RF power [6 min] = 2 J/kg; Whole body averaged power = 0.2 W) and EGG measurements were performed in a 34y-old female singer during different phonatory tasks. As a precaution, the pulse sequence used low flip angles (α = 6°), and optical temperature probes were attached to the EGG electrodes. TrueFISP lung images were obtained with a combination of a spine matrix coil and a body matrix coil. To evaluate the performance of the EGG system, the volunteer was asked to sing different pitches (A3, A4, and A5) for 10 s, during which the first 5 s were acquired without a simultaneous MRI image acquisition, and the subsequent 5 s during lung MRI. The un-distorted EGG signal and post-processed EGG data during MRI were then analyzed using the EGG software package SPEAD (Version 4.2.1, Laryngograph Ltd, Walligton, UK). The following standard EGG parameters were calculated: fundamental frequency, 50% contact quotient, frequency jitter and shimmer.

In a second phonatory task the volunteer was asked to start singing at high pitch (A5, 880 Hz) followed by a rapid double octave jump to A3 (220 Hz). In this task, the fundamental frequency as measured with EGG was plotted together with maximal extension of the right lung (i.e. lung diameter) which was manually extracted from the coronal trueFISP images as a surrogate measure of lung volume. The rates of exhalation during singing of A5 and A3 were calculated for the corresponding portions of the lung diameter curve by a linear fit.

As a reference, an MR compatible calibrated microphone (Prepolarized Freefield 1/2" Microphone Type 4189, Bruel&Kjær, Nærum, Denmark) was included in the experiment setup. The signal voltage of the microphone transducer that is carried along a coaxial cable fed through the Faraday's cage with a BNC connector was then recorded via an oscilloscope outside of the magnet room. The sound data was processed with Matlab®.

4 Results

4.1 MRI of Samples with Ultra-short Transverse Relaxation Times

4.1.1 Sequence Comparison and Parametric Imaging Results

In all MR images tissue differentiation was feasible, and major anatomical structures such as bones, tendons and muscles could be clearly identified (Fig. 50). Skin soaked with embalming resin has a very high signal intensity in all MR sequences, whereas it appears iso-intense to the neighboring tissues in CT. PETRA offers the highest SNR for all tissues followed by UTE, and SPI has the lowest SNR efficiency (Table 1). Due to longer acquisition delay (i.e. effective TE), SPI has a lower SNR especially for anatomical structures with shorter T_2^* values. Note, that SPI SNR values were not normalized. For an appropriate parameter scaling with respect to the voxel sizes, sequence parameters and the total acquisition times SNR values for 1.5T and 3T SPI should be scaled to 0.09 and 0.1, respectively. With the increasing field strength from 1.5 T to 3 T, SNR increases by a factor of about 2.

It is observed that the tissue with longer T_2^* have higher SNR as expected. SPI has lower SNR per unit time while UTE and PETRA has comparable SNR per unit time, which could be seen at the sequence parameter scaling. The lower SNR of SPI is also due to the longer acquisition delay which is enforced by the limitations of the gradient systems in use.

		SNR					Relaxation Times			
	Anatomical structure	UTE		PET-RA	SPI		T_1 (ms)		T_2^* (μs)	
		1.5T	3T	1.5T	1.5T	3T	1.5T	3T	1.5T	3T
2	Flexor pollicus longus tendons	16	32	21	7	18	182±38	78±11	214±29	158±31
3	Flexor digito-rum profundi tendons	14	44	24	6	37	206±42	75±12	321±36	182±32
4	Dorsal interos-sei muscles	15	22	21	6	17	193±43	81±14	315±44	248±33
5a	Dorsal skin w/ e. resin*	27	51	43	19	47	268±43	88±17	458±76	435±56
5b	Palmar skin w/ e. resin	51	83	61	40	81	266±41	89±16	417±34	544±70
5c	Ulnar skin w/ e. resin	68	127	72	71	76	263±53	74±12	526±88	565±79
Ia	Metacarpal I (sb**)	4	8	8	1	6	NA	85±17	NA	143±17
IIa	Metacarpal II (sb)	8	19	17	2	10	NA	105±16	NA	152±23
IIIa	Metacarpal III (sb)	5	20	17	3	15	NA	93±18	NA	138±16
IVa	Metacarpal IV (sb)	6	16	16	2	16	162±45	97±13	119±20	131±20
Va	Metacarpal V (sb)	6	18	17	2	14	188±30	94±15	115±14	147±28
Ib	Metacarpal I (cb***)	12	26	21	9	32	216±53	73±17	191±18	212±35
IIb	Metacarpal II (cb)	12	29	26	10	34	212±49	72±18	209±19	215±31
IIIb	Metacarpal III (cb)	19	36	31	12	35	235±38	80±15	168±22	173±18
IVb	Metacarpal IV (cb)	13	30	26	10	32	233±44	77±17	227±26	207±21
Vb	Metacarpal V (cb)	11	30	30	11	27	211±43	75±19	178±19	228±34

Tab. 1: Image SNR and relaxation values of the labeled anatomical regions for 1.5 T and 3 T. *w/ e. resin: with embalming resin. **sb: spongy bone. ***cb: cortical bone.

In Table 1, the relaxation values of the corresponding anatomical regions are given. T_1 and T_2^* maps for both magnetic fields are shown in Fig. 51. T_2^*

varies between 90 - 600 μs and T_1 varies between 60 - 300 ms. Metacarpal spongy bone has very low SNR, and here relaxation values could not be calculated. T_2* of skin soaked with embalming resin is significantly higher than that of other tissues, whereas metacarpal spongy bone has the lowest T_2*. For tendons T_2* decreases with increasing field strength; however, T_2* of skin soaked with embalming resin and the metacarpal bones does not change with the field strength. T_1 values do not differ significantly among the tissues - only skin soaked with embalming resin has higher T_1 in 1.5 T measurements. T_1 values decrease by a factor of two with increasing field strength.

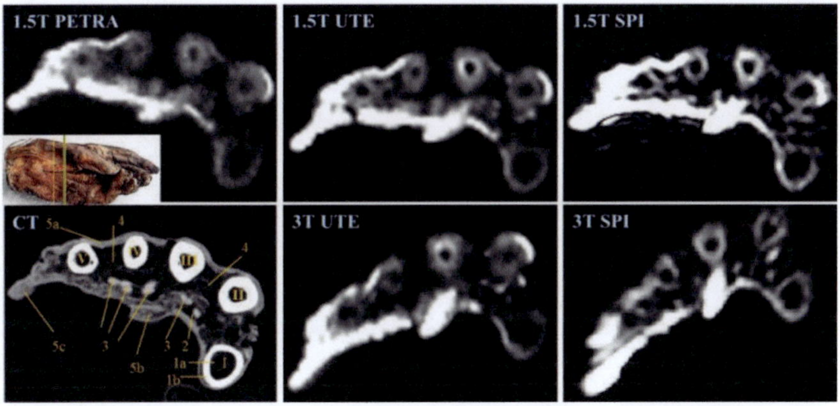

Fig. 50: CT and 1.5T and 3T MR images corresponding to the slice shown with the yellow line. Significant anatomical structures are labeled on the CT image. See Table 1 for descriptions and local SNR calculations, along with the relaxation time values.

In the images (Fig. 50) small anatomical structures are not clearly visualized in the PETRA data due to image blurring, whereas in UTE edges appear sharper (for example, at the index finger). The source of blurring for UTE and PETRA is the fast decay of the transverse magnetization. For SPI, blurring is observed especially at 3T field strength; because the point spread function is disturbed due to the RF pulse being applied while the gradients are switched on.

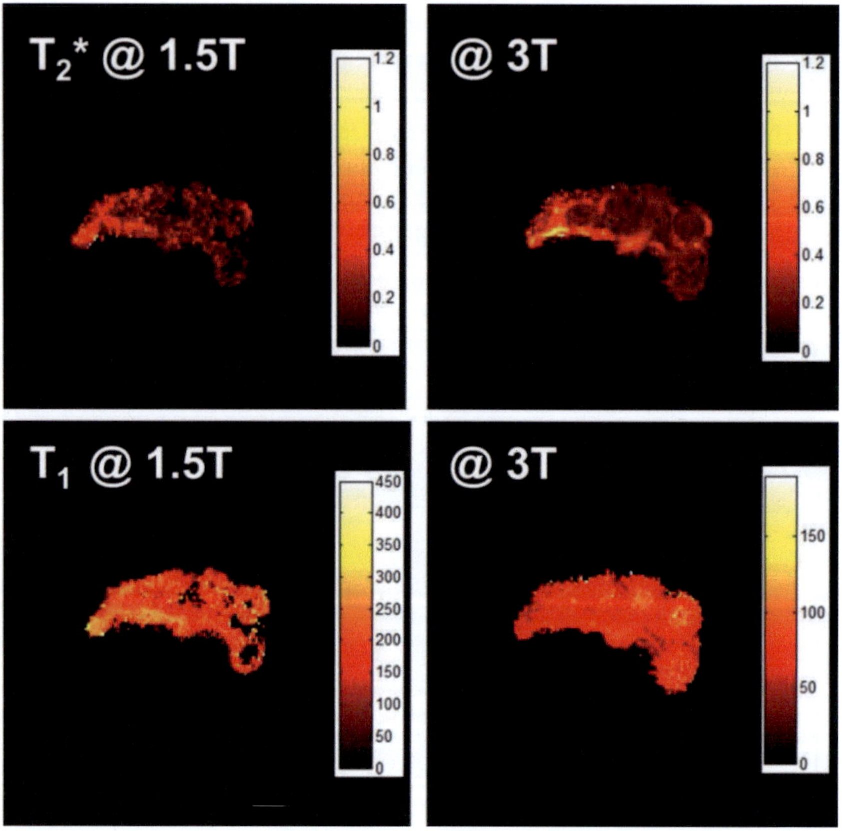

Fig. 51: T_2* relaxation map in ms for 1.5T (upper left) and for 3T (upper right). T_1 relaxation map in ms for 1.5T (bottom left) and for 3T (bottom right).

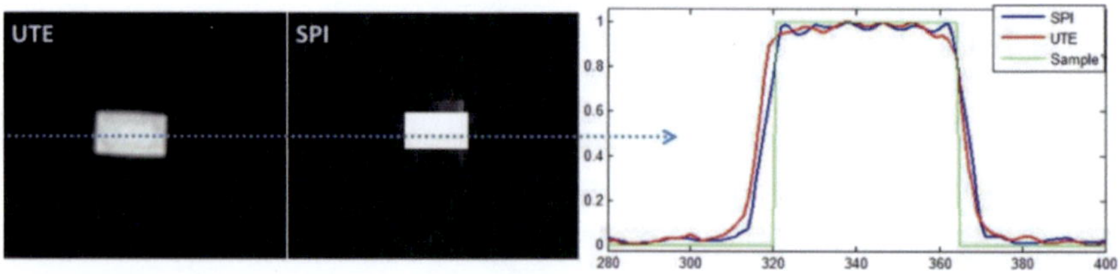

Fig. 52: Line profile comparison (normalized signal intensity vs. pixel number) for UTE and SPI on a rubber phantom with rectangular cross-section.

In Fig. 52, UTE and SPI images of a rubber with rectangular cross-section are shown. It is observed from the line profiles that SPI preserves the sharp edge information better than UTE. The blurring effect becomes more evident with decreasing T_2*.

4.1.2 Modifications to the Short-TE Pulse Sequences

4.1.2.1 Gradient Switching Corrections in UTE Sequence

In Fig. 53, 2D UTE images of the resolution phantom are shown when reconstructed with and without gradient delay correction. Estimated gradient delays were 1.3 μs, 2.2 μs in x and y directions, respectively. Gradient delay correction using opposite spoke cross-correlation helped reducing the blurring and signal shift artifacts. Image distortions at the edges were significantly corrected.

Fig. 53: A structural image of the resolution phantom using a standard GRE sequence (left). UTE images of the resolution phantom reconstructed without (middle) and with (right) gradient delay correction.

4.1.2.2 Gradient friendly modifications in SPI

The temperature measurements during sequence execution shows that with standard step-wise k-space ordering (Seq. 1), higher temperatures are measured than with the reordered sequences 2 and 3 (Fig. 54). While sequences 1 and 2 could not be executed by the MR system due to a cooling system warning, Seq. 3 could finish data acquisition without interruption. Gradient waveforms for three directions in Seq. 3 are similar as well as the measured increase in temperature. Therefore, the modified sequence can be performed with more extreme parameters compared to standard version. Results also show that gradient coil heating not only depends on the instant current levels but also on k-space ordering since the cooling system has a non-negligible time constant.

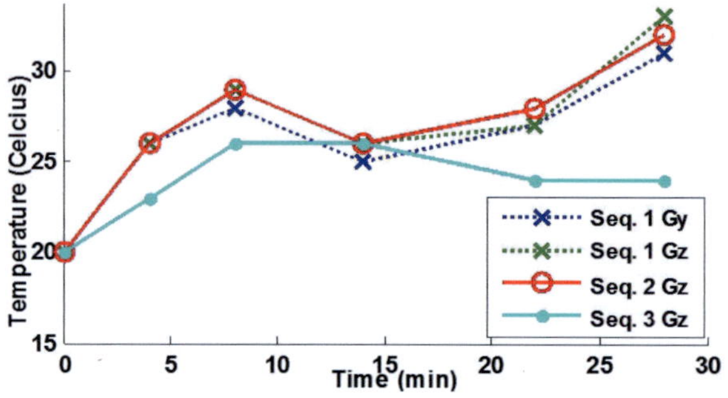

Fig. 54: Temperature measurements for four different gradient waveforms described in Methods, Fig.32. Measured temperature increase for the modified saddle-like k-space trajectory is substantially lower at the end of the measurement, and unlike the other gradient waveforms, it does not increase monotonically after 15 min.

Bloch simulations show that the slice profile improves with the limited gradient amplitude, and even at the edges of the FOV an excitation of 80% is achieved. In Fig. 55 one can see that the limited gradient amplitudes provide a smoother signal distribution over FOV. The additional effect of the gradient limitation on gradient heating was minor for TR = 6.4 ms.

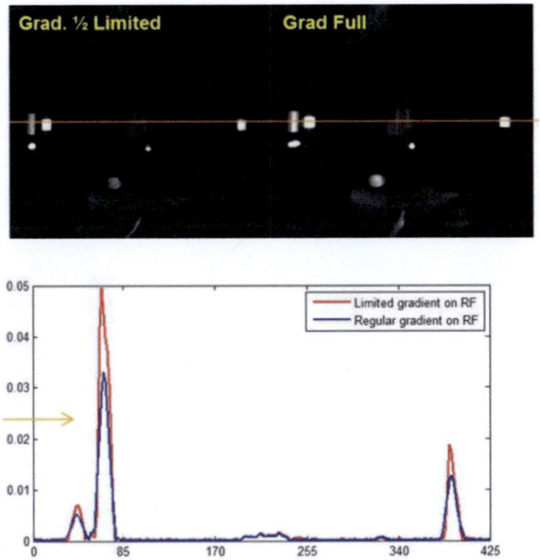

Fig. 55: Comparison of sequences with regular gradient level (top-right) and gradient level halved on RF (top-left) using a large loop resonator. Slice profile is smoother when the gradient level is halved during excitation, thus the edges are not suppressed (bottom).

4.1.3 Hardware Optimization for MRI of Ultra-short T_2 Samples

4.1.3.1 RF Resonators

Fig. 56 shows the homogeneity comparison of three resonators of similar dimensions. Fig. 56a,b,c shows the B_1 maps generated with an SNR mask and scaled to between 30 μT and 35 μT. Along the radius

$|B_1|$ varied by 3.2%/ 2.8% (1.5 T / 3 T) for the solenoid resonator. And 1.6% and 3.9% for the loop-gap and birdcage resonators, respectively. Fig. 56d,e,f shows SNR maps for the structural UTE images of the mummified hand at 3T acquired with each of the resonators for the same flip angle. In Fig. 56g,h,i, SNR histograms are plotted. Loop-gap resonators had a higher homogeneity, and more pixels with a SNR of greater than 100. Homogeneities of birdcage and solenoid resonators were comparable, yet solenoid outperformed the birdcage in terms of SNR.

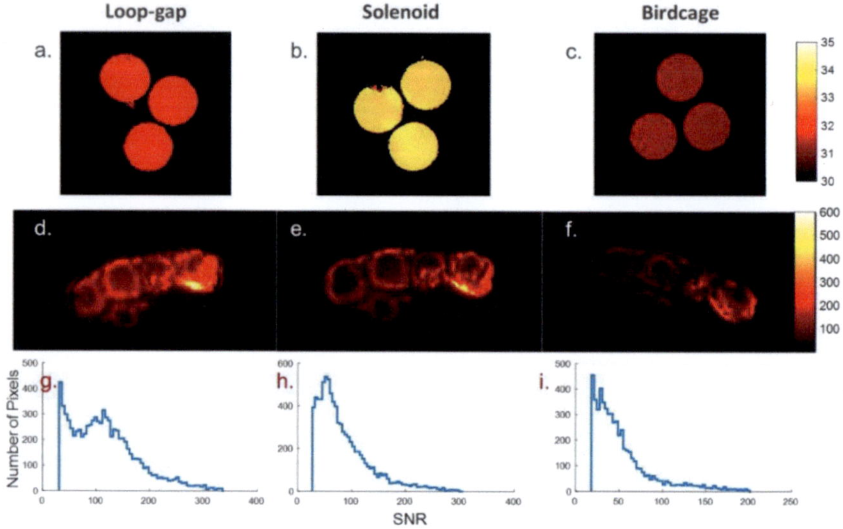

Fig. 56: B_1 field maps when loaded with distilled water tubes described in section 3.1.3. for loop-gap (a.), solenoid (b.), and birdcage (c.) resonators. UTE images of the mummy hand acquired with each resonator (d.), (e.), (f.), and corresponding SNR histograms (g.), (h.), and (i.), respectively.

4.1.3.2 RF Circuitry: Tx/Rx Switches

The linear Tx/Rx switches provide up to 50 dB isolation. Switching times of 2.4 μs / 3.6 μs were measured for 1.5 T / 3 T linear Tx/Rx switches. Isolation and switching times for the quadrature Tx/Rx switch were 42 dB and 2.1 μs, respectively.

4.2 Concurrent Excitation and Acquisition

4.2.1 Active Decoupling of RF Resonators using a Transmit Array System

4.2.1.1 Decoupling of RF Resonators with Additional Decoupling Resonators

Geometric decoupling reduced the induced voltages in the Rx resonator from 80 ± 5 mV_{pp} to 8 ± 1 mV_{pp}, and PA decoupling further decreased the voltage to less than 10 μV_{pp}. Thus, with the geometric decoupling alone, an isolation of 18±3 dB was achieved, and PA decoupling additionally yielded an isolation of 52±5 dB. In total, a decoupling of 70±8 dB was possible with the proposed setup.

For chirp pulses PA decoupling deviated 10 dB over 30 kHz band. The resulting plot of the geometric decoupling is shown in Fig. 57 and for the PA decoupling using the multichannel transmission system in Fig. 58. Voltage levels and decoupling values are given as an average over 60 repeated measurements.

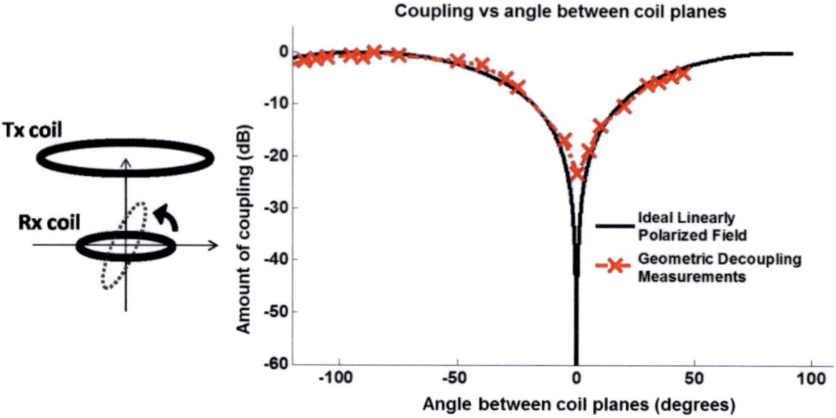

Fig. 57: Measured B_1-induced voltage for a rotation of the Rx resonator with respect to one of the Tx resonators. An ideal linearly polarized field is also calculated, and the angle between the resonators for both fields are scaled around 0°.

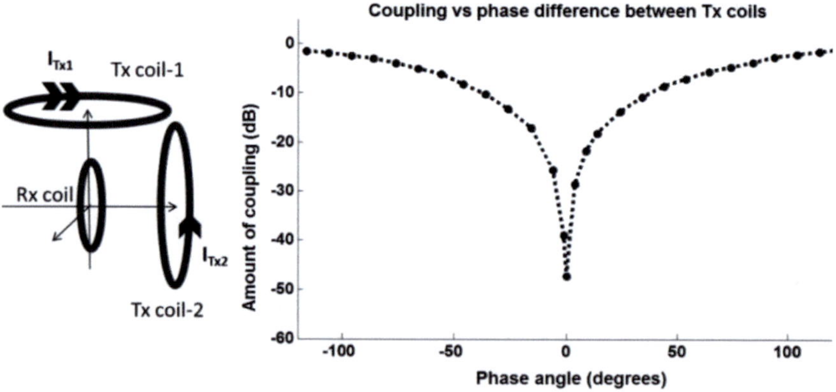

Fig. 58: Decoupling vs. phase of an RF generator unit. Phase and amplitude (PA) decoupling is represented by changing the phase of one Tx channel step by step over one cycle. The data are scaled around 0°.

From the B_1 mapping experiment, the B_1 magnitude of the rectangular RF pulse in the CEA experiment was calculated to 0.1 μT for rectangular RF pulse of 8 ms duration and U_{peak} = 1 V_{pp}. Note that B_1 was calculated only for the Tx coil-1.

4.2.1.2 Transmit Noise

Without geometric decoupling the standard deviation of the acquired CEA signal was 14 μV; with 10 dB and 20 dB geometric decoupling, the standard deviation was 11 μV and 3 μV, respectively. The acquired signal voltage levels were the same when the transmit system was driven with an RF pulse with zero amplitude, and when the transmit system was turned off and unplugged. Therefore, it was concluded that the transmit noise-induced voltage does not exceed the receive noise floor which is measured as 5 μV_{pp}.

4.2.1.3 Application to MRI with CEA

Fig. 59 shows the acquired raw data processed through leakage subtraction, a smoothing filter, and deconvolution. Contamination from B_1-induced leakage signal can be observed as a frequency dependent DC offset in the raw data with the sample (Fig. 59a). B_1-induced leakage ϑ_{leak} is measured by repeating the same acquisition after the sample is removed (Fig. 59b). Subtraction of the waveforms results in the signal waveform without contamination as is shown in Fig. 59c. The remaining DC offset in the data was also subtracted before the next step. Deconvolution was then performed in the Fourier domain using the analytical expression for the chirp pulse $B_1(t)$ from Eq. 2.4.14 as the deconvolution kernel yielding a radial projection of the data (Fig. 59d). For gridding reconstruction, an inverse Fourier transform was performed to obtain the frequency-encoded FID signal (Methods, Fig. 59e).

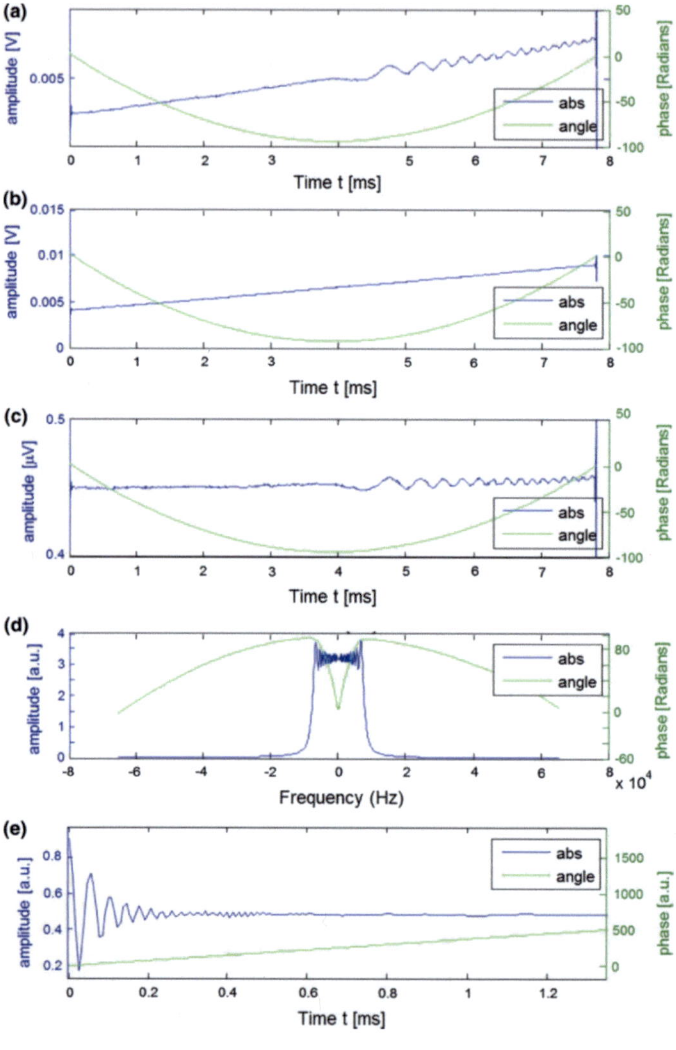

Fig. 59: (a) Raw data acquired from a water sample in the presence of a 10 mT/m gradient field along the x direction: $\boldsymbol{s_{Rx}(t) = h(t) \otimes b(t) + \vartheta_{leak}(t)}$. (b) Measured input B_1 waveform as a chirp function representing a leakage signal that is convolved by the system's response function: $\boldsymbol{\vartheta_{leak} = A_{cpl}\vartheta_{Tx}e^{j\varphi_{cpl}}}$. (c) Acquired signal after the leakage subtraction: $\boldsymbol{s(t) = h(t) \otimes b(t)}$. (d) Fourier transform of the estimated FID signal after deconvolution: $\boldsymbol{H(\omega) = S(\omega)B^{*}(\omega)/|B(\omega)|^{2}}$. (e) Single-sided FID signal, which is an approximation of the actual FID: $\boldsymbol{\widetilde{h}(t) = \mathcal{F}(H(\omega))}$.

The reconstructed CEA image after re-gridding is shown in Fig. 60. The rubber phantom with all its inner structures is clearly visible, and some blurring is observed at the edges of the phantom.

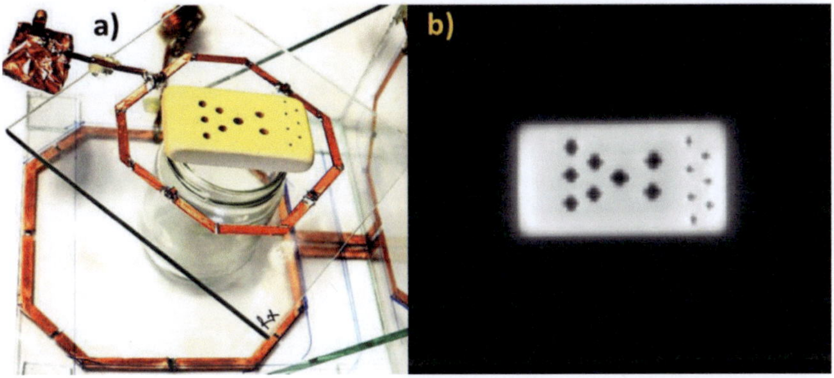

Fig. 60: (a) Rubber phantom with a T_2* value of 500 µs with holes, (b) CEA image with a base resolution of 256 and 768 radial spokes; TR= 200 ms. There is blurring at the edges, but the holes of various dimensions (0.1-0.4 mm) drilled on the rubber are clearly visible.

Center slices of 3D GRE (FLASH) and CEA data are shown for three orthogonal planes in Fig. 61. The anatomical details are available in GRE images. The artifacts and the blurring together results in loss of significant anatomical details, yet the major structures are visible in the first proof-of-principle 3D CEA image slice. The left kidney of the mouse is pointed and visible in both images. The contrast properties of the CEA image is needed to be further analyzed.

An instability of the RF transmit system is observed even with the achieved decoupling during RF excitation in Fig. 62. In acquisition of each spoke, small deviations in the B_1-induced leakage signal are present.

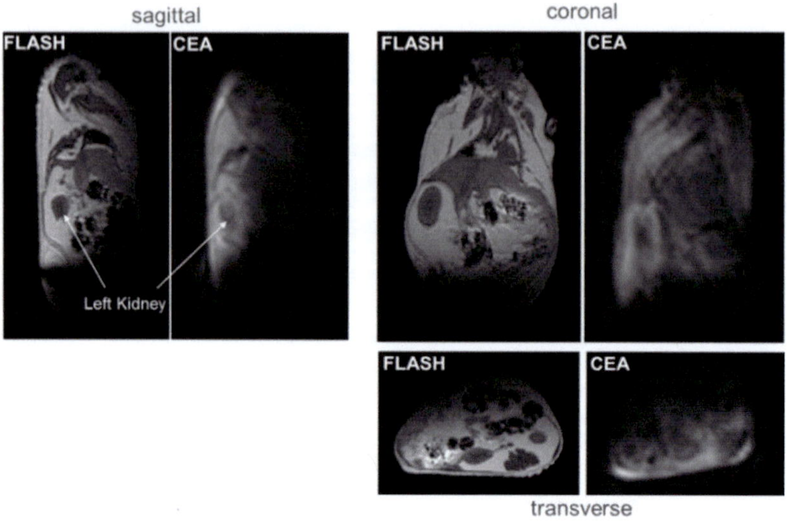

Fig. 61: Sagittal, coronal, and transverse slices from 3D MRI of ex vivo APOE mouse with CEA and GRE (FLASH). GRE image shows all the anatomical details with 0.4 mm in-plane resolution. Major structures are visible in CEA image such as the left kidney.

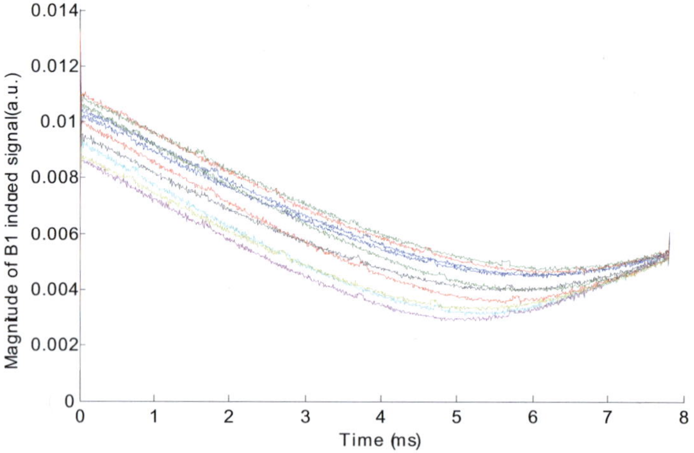

Fig. 62: Variation of B_1-induced leakage signal over subsequent measurements after decoupling. Due to the deviations in the leakage signal, the resulting images are prone to artifacts since the leakage signal is assumed to be constant during the acquisition of the entire k-space.

4.2.2 Automated Analog Cancellation using Dynamic Feedback from MRI System

4.2.2.1 Analog Cancellation Performance tests

The stability and noise performance of the analog cancellation system are shown in Fig. 63. The temporal signal variation in the unloaded resonator amounted to 0.2% of the mean of the signal. In the noise measurement the receive noise increased by up to 9.7% from the noise floor when the Tx resonator was operated with 0 V input.

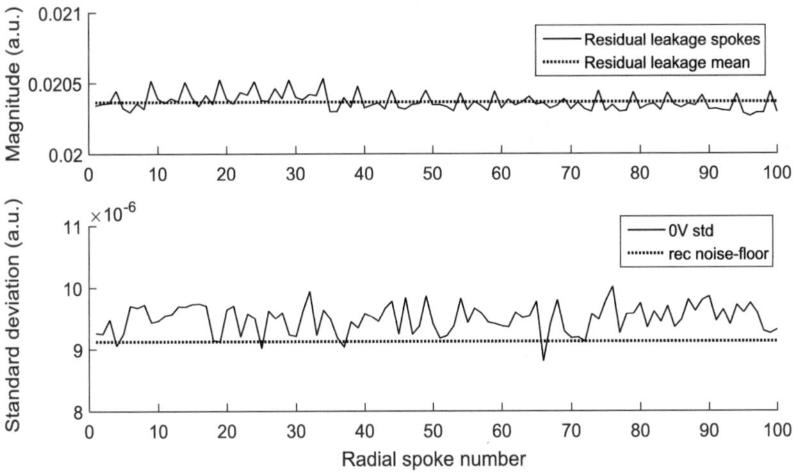

Fig. 63: Stability and noise checks of the analog cancellation system during MRI. Upper part shows the difference between the residual leakage of two spokes and the mean of all the spokes for maximum and minimum cases. Maximum of 0.2% magnitude deviation from the mean was observed. At the middle of the sweep, MR signal dependent modulations starts to be seen. Bottom part shows the noise standard deviations for each spoke compared to the mean of all spokes and the receive noise floor. Maximum of 9.7% increase from the receive noise floor was observed.

In Fig. 64 comparison of acoustic pressure measurements are shown. Background sound level (i.e. the default function of the Helium-pump) was measured as 49.4 ± 0.1 dBA. CEA sequence with 60000 spokes performed at 54.8 ± 0.3 dBA for the available maximum gradient amplitude range. UTE sequence with the same number of spokes performed at 78.6 ± 2.0 dBA for maximum gradient strengths G_{max} = 20 mT/m, and at 74.1 ± 1.8 dBA for G_{max} = 10 mT/m. The GRE sequence with the reported parameters (cf. section 3.2.2.4.) performed at 81.8 ± 0.4 dBA. Volunteer measurement also confirmed the low noise behavior of the pulse sequence, and the volunteer didn't need to wear any ear protection during the measurement. During the CEA sequence, a low-volume whistling sound was present.

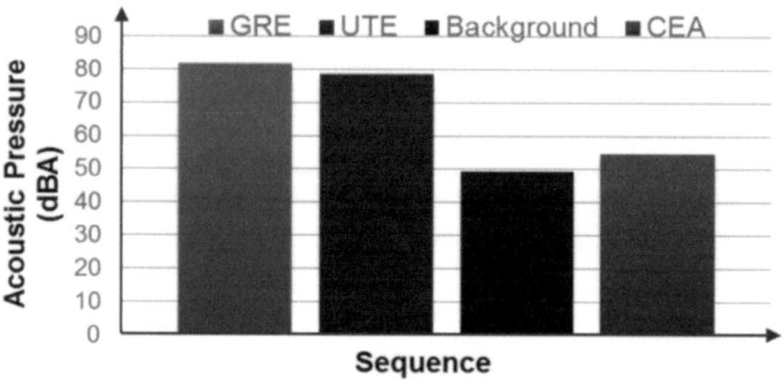

Fig. 64: Acoustic noise performance results. Continuous gradient-ramps in CEA reduced the acoustic noise significantly, close to the system's background noise.

4.2.2.2 Application to MRI with CEA

In Fig. 65 a comparison is shown of a coronal slice of the phantom acquired with CEA and GRE. The phantom tubes are clearly visible in both acquisitions (bottom), but the surrounding rubber (top) has 3.5 fold higher signal-to-noise ratio with CEA than with GRE. In CEA image, artifacts due to radial sampling are present due to system imperfections such as time-dependent leakage signal. Another source of the artifacts is the non-Cartesian reconstruction, which is sensitive to the small deviations in the sampling space.

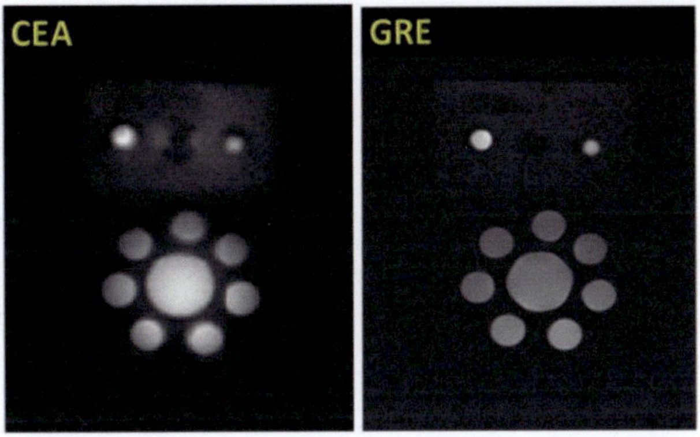

Fig. 65: Coronal slice from the 3D CEA MR image of the tube and rubber phantoms (left) and GRE image as the structural reference.

Fig. 66 shows a coronal slice of the 3D data of the human wrist acquired with CEA and GRE. Slices are not identical since the experiments are conducted at different times. The wrist bones are seen with a higher intensity in CEA image, and extensor finger tendons, which are not represented in the GRE image, provide higher signal intensity in CEA than in GRE. Carpal bones also have a higher signal in CEA image even in trabecular bone. Overall contrast to noise is higher in GRE due to intrinsic T_2* weighting.

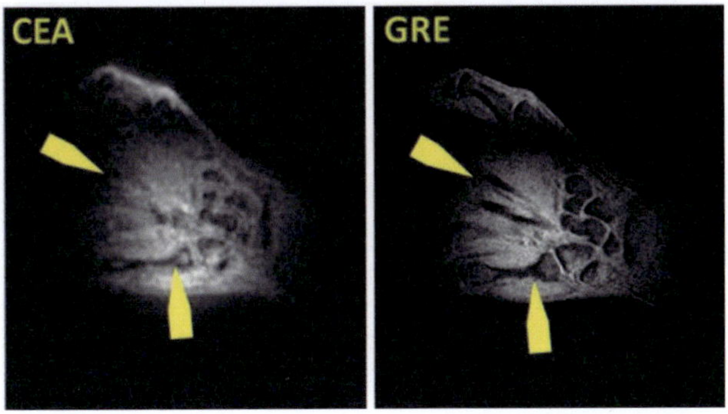

Fig. 66: Coronal slice from the 3D CEA MR image of the healthy volunteer wrist (left) and GRE image as the anatomical reference.

4.2.2.3 T_1 Modulation Effects in Frequency Swept MRI with CEA

In Fig. 67, TR_{eff} for the three different trajectories and the resulting modulations in M_{xy} are shown using a 2D sequence with $\alpha = 10°$ for a point source with $T_1 = 100$ ms placed at (x,y) = (150,150) on a 512x512 grid. In Table 2, resulting amount of change in TR_{eff}, M_{xy}, and the PSF for appoint source at the edge of the field of view (FOV) are summarized for all 2D and 3D trajectories for T_1 values of 10 and 100 ms and α values of 10° and 30°.

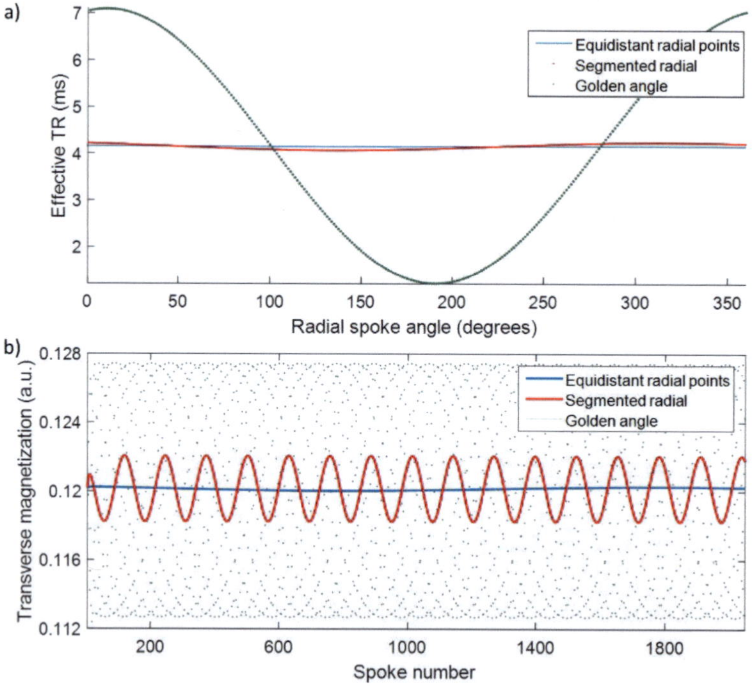

Fig. 67: Effective TR, TR_{eff} with respect to the radial spoke angle for three different 2D k-space trajectories for a sample located in (x,y) = (150,150) on a 512x512 grid (a). Modulation in transverse magnetization with respect to the radial spoke number (b). The most extreme modulations are observed with golden angle ordering.

T_1 modulation effects for sequential ordering of equidistant points were below 0.2% for all conditions. However, with segmented ordering, changes up to 1.8% are calculated in PSF. Golden angle trajectory resulted in the most extreme modulations in effective TR, transverse magnetization and PSF for majority of the simulated conditions. The maximum change in PSF of 5.2% was observed for 2D golden angle trajectory (Fig. 68).

Sequence	Trajectory	ΔTR_{eff} (%)	T_1 (ms)	Flip angle (α)	ΔM_{xy} (%)	ΔPSF (%)
2D	Equidistant points	0.44	100	10°	0.088	0.04
				30°	0.094	0.12
			10	10°	0.015	0.005
				30°	0.097	0.038
	Segmented	7.03	100	10°	1.086	0.41
				30°	1.5	1.8
			10	10°	0.24	0.085
				30°	1.54	0.61
	Golden angle	189	100	10°	1.8	0.6
				30°	5.7	5.6
			10	10°	2.04	0.56
				30°	16	5.2
3D	Equidistant points	7	100	10°	0.11	0.02
				30°	0.27	0.09
			10	10°	0.09	0.02
				30°	0.69	0.11
	Segmented	63	100	10°	0.8	0.41
				30°	2.5	1.2
			10	10°	0.74	0.18
				30°	5.6	1.5
	Golden angle	60	100	10°	0.8	0.12
				30°	2.5	1.2
			10	10°	0.77	0.37
				30°	5.9	1.1

Tab. 2: Simulation results for comparison of three k-space ordering strategies in 2D and 3D. The further the consecutive spokes are apart from each other, the higher the %change in TR_{eff}, M_{xy}, and PSF.

As a result of sequentially satisfying the resonance conditions in frequency-swept pulses, the k-space representation of the signal is also affected. Although in most of the cases the PSF-related signal change is less than 5%, modulations can cause blurring artifacts under extreme conditions (radial segmentation schemes with lines far apart, short T_1, and/or high α). To avoid PSF artifacts, readout gradients should preferably be adjusted to follow sequential ordering of equidistant points in k-space, and low flip angles are more preferable.

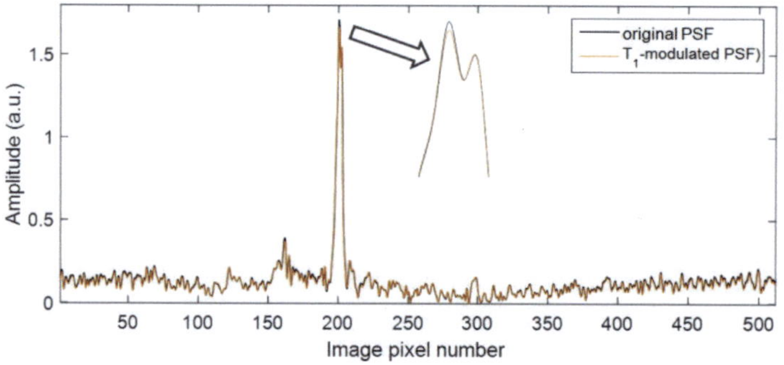

Fig. 68: Comparison of PSFs for T_1-modulated and constant TR reconstruction for a point source at the edge of the FOV (x,y) = (200,200) on 512x512 grid. Here the case with maximum of 5.2% difference in PSF is shown, which was calculated for golden angle ordering.

4.3 EGG Measurements during Dynamic Lung MRI

4.3.1 EGG Device

RF shielding efficiency tests resulted in noise standard deviation values of 3.35, 3.41, and 3.46 corresponding to the 'no device', 'device off' and ' device running' cases. No RF spikes were visible in the noise images. The shielding effectively prevented any RF interference from being carried inside the magnet room.

4.3.2 Signal Processing

Fig. 69 shows the measured EGG signals during constant singing of the note A5 (EGG_{ideal}), during trueFISP imaging and singing (EGG_{meas}), and the RF induced voltage signal during trueFISP imaging alone ($RF_{coupled}$). For comparison, the RF pulse measured with a small B1 probe is shown (RF_{in}) together with the corrected EGG signal after retrospective processing by ICA (EGG_{corr}). Note, the EGG_{ideal} signal was measured to be identical inside the magnet bore and outside the magnet room, no obvious distortion due to the magnet field was observed. As can be seen in the EGG_{meas} signal, flip angle values were set low enough to prevent the receive circuitry of the EGG device from being saturated. The applied RF pulse for trueFISP sequence has a sinc shape and the amplitude modulation of the original RF pulse was mainly preserved after being coupled into the electrodes of the EGG device. RF-induced voltages occurred exactly at each TR as expected. In the corrected EGG signals some remaining contamination from the RF pulse is visible.

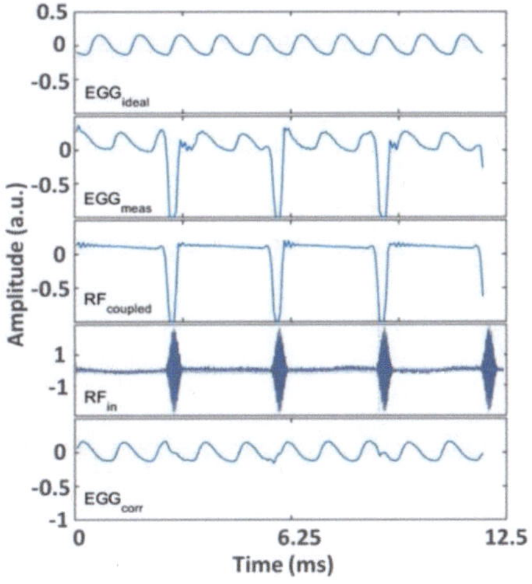

Fig. 69: Comparison of EGG signals for constant singing of A5: the ideal EGG signal with no RF pulse is applied (EGG_{ideal}), EGG signal during 2D TrueFISP pulse sequence (EGG_{meas}), RF induced voltage signal no singing ($RF_{coupled}$), the applied RF pulse measured by oscilloscope (RF_{in}) and the EGG signal after retrospective processing (EGG_{corr}).

Table 3 shows the calculated EGG signal parameters during singing of the notes A5, A4, and A3. Comparison of the EGG parameters between EGG_{ideal} and EGG_{corr} (Table 3) shows that the remaining signal distortion after ICA correction only minimally changes the values of fundamental frequency (by 1%), 50% contact quotient (by 4%), jitter (2%), positive (3%) and negative shimmer (4%).

Pitch	A5		A4		A3	
	without RF	with RF	without RF	with RF	without RF	with RF
Fundamental frequency (Hz)	857	851	440	436	222	224
Contact quotient	47%	46%	41%	45%	44%	44%
Jitter factor	3%	4%	2%	3%	1%	2%
Shimmer+	1%	0%	2%	0%	3%	0%
Shimmer-	-2%	0%	-2%	0%	-4%	0%

Tab. 3: Comparison of vocal fold characteristics for singing A5, A4 and A3 pitches without MRI sequence running and during trueFISP sequence (cf. section 3.3.4.).

4.3.3 MR Safety Tests

In the safety tests, no displacement or rotation was observed in the electrodes due to the static magnetic field. In the SE and GRE images the typical image artifacts were observed in the vicinity of the electrodes. Artifact widths measured from the center of the electrodes as shown in the transversal slices orthogonal to the electrodes were maximum 43 mm and 45 mm for SE and GRE images, respectively (Fig. 70a).

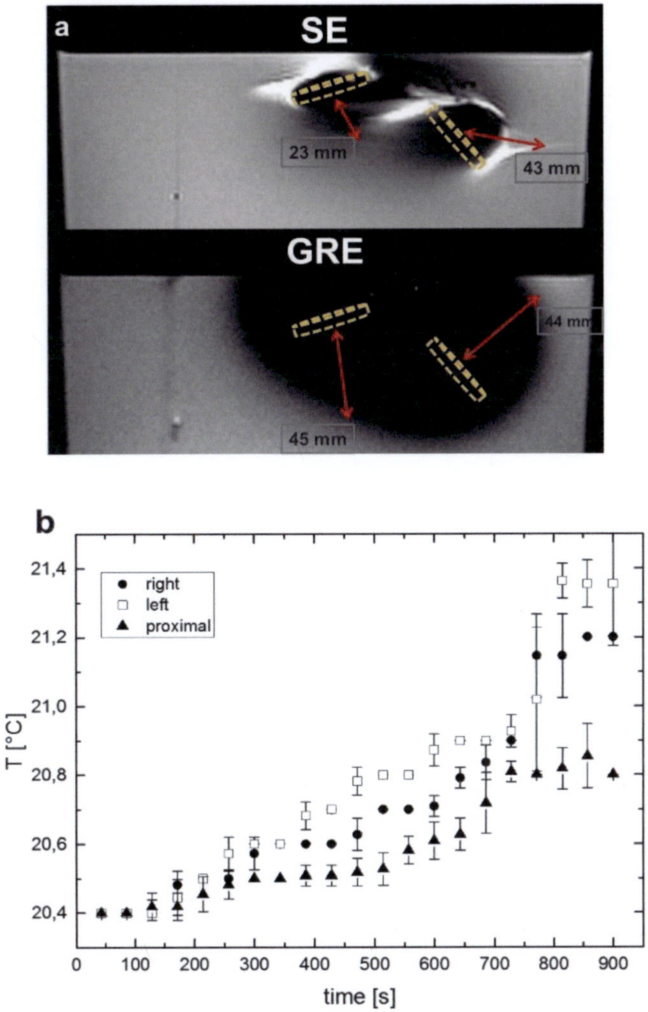

Fig. 70: (a) SE and GRE sequences were used for artifact width measurements. The edge of the container was used as reference for calculating the exact position of the electrodes. (b) Temperature measurement results over 15 min of RF transmission with TR = 30 ms. The temperature probes attached to the surface of the electrodes display higher temperature rises. Proximal probes was placed in close proximity (distance larger than 1 cm) of the electrodes.

The electrodes produced artifact within a distance of 51 mm from their edges. Since imaging FOV is far away from the position of the electrodes, no image artifacts was present (Fig. 71) due to electrodes or With the high-SAR protocol a temperature increase of $\Delta T < 1.2°C$ was measured in the phantom (Fig. 70b). Temperature measurements over 15 min show that the highest temperature increase is found at probes 1 and 2 at the electrode surface, whereas the temperature probes placed at distances larger than 1 cm measured 0.6°C increase or less. A reference probe at a distance larger than 15 cm displayed no temperature elevation over the experiment. From the temperature increase a maximum local SAR value of 5.53 W/kg was calculated. This results indicate that EGG device can be safely used in pulmonary MRI measurements at 1.5 T field strength, under maximum slew rate of 142 mT/m/s or less, and maximum MR system reported whole body averaged SAR of 2.1 W/kg for 15 min of scanning.

4.3.4 Volunteer Measurements

During the volunteer measurements the optical temperature probes at the electrodes displayed a $\Delta T = 0.4°C$ increase at the end of the total experiment duration. Changes of lung volume during singing could be observed simultaneously with the EGG recording, and subjects did not report any sensations caused by the presence of the EGG electrodes and the cables. As a surrogate marker for the lung volume, right lung diameter during singing is plotted in Fig. 71b together with the fundamental singing frequency measured with the EGG. After deep inhalation, the subject started singing at a high pitch (A5). With a glissando (without any breaks or breathing in) a two-octaves jump to a lower pitch (A3) is performed. The whole procedure was repeated three times during the total dynamic image acquisition time of 42 sec at approximately 3 fps. Slopes of the linear fits for lung diameter curve portions during singing of A5 and A3 are -7.9 and -3.6, respectively for the first performance; -9.7 and -4.6 for the second, and -16.3 and -4.4 for the last repetition. Prior to the two-octaves jump, the lung diameter is increased by 3.4 mm, 5.5 mm, and 8.3 mm for the first (Fig. 71a), second and the third repetition of the same task.cables.

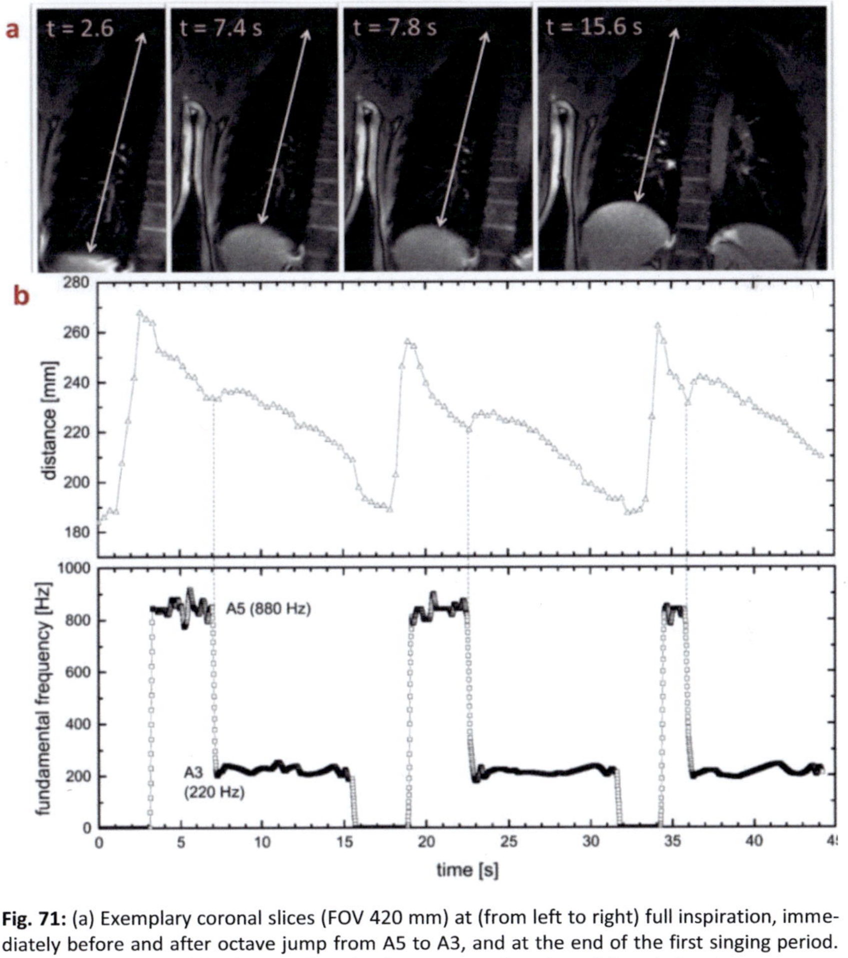

Fig. 71: (a) Exemplary coronal slices (FOV 420 mm) at (from left to right) full inspiration, immediately before and after octave jump from A5 to A3, and at the end of the first singing period. (b) Lung diameter and fundamental singing frequency as function of time during three consecutive singing periods with octave jumps. Lung diameter was manually obtained from the 2D trueFISP images as the distance between diaphragm and top of the right lung. Fundamental frequency was calculated from simultaneously acquired EGG data.

5 Discussion and Outlook

5.1 MRI of Samples with Ultra-short Transverse Relaxation Times

5.1.1 Sequence Comparison in MRI of Ancient Remains

In this work a comparison between three ultra-short TE imaging techniques was presented for MR imaging of ancient mummified remains in clinical MRI systems. The imaging pulse sequences use different k-space acquisition modules, which either scan k-space radially, individually from point to point, or through a combination of both.

The SNR analysis of the sequences shows that SPI has the lowest SNR followed by UTE and PETRA, whereas SPI gives the sharpest images followed by PETRA and UTE. Thus, PETRA is the preferable imaging sequence in situations where image sharpness is less important than signal intensity or signal contrast, as e.g. in objects with a low proton spin density such as bone. SPI, on the other hand, is the best imaging technique if spatial resolution is of importance, as e.g. in the visualization of small subcutaneous layers. Finally, UTE can be used when T_2^* is on the order of or longer than the readout time. In this case, UTE allows acquiring data sets more efficiently than SPI, but it suffers from blurring caused by the signal decay during data acquisition. Considering the diversity of ancient mummified samples, the results of this study cover only a certain domain of dehydrated mummies of Ancient Egypt, and more ancient mummified samples need to be studied to define standard protocols for MRI in this field.

With the solenoid resonator, in majority (97% at 1.5T; 100% at 3T) of all anatomical regions T_1 and T_2* values could be measured. T_1 values of the mummified hand are only slightly smaller than typical T_1 values found in *in vivo* studies at the same field strength. T_2* values, however, are more than one order of magnitude smaller than in humans. Due to the applied threshold masks, the T_2* relaxation maps do not indicate all anatomical structures. Especially for metacarpal spongy bone, the SNR was too low to calculate reasonable relaxation time values due to low spin density. Here, lower acquisition delays (i.e., shorter TEs) and more sensitive resonator designs would be beneficial to achieve a higher SNR.

The ultra-short TE sequences require a high fidelity of the imaging hardware and the use of Tx/Rx resonators. In particular, the dead times between the end of RF excitation and the beginning of data acquisition must be minimized. Local Tx/Rx resonators perform better in terms of SNR, acquisition delay, and power efficiency than the standard clinical setup using a large Tx body resonator combined with additional Rx only resonators.

At higher field strength most of the tissues had shorter T_2* values, which is in line with the simple model of an apparent transverse relaxation rate $1/T_2$* that is inversely proportional to the field inhomogeneities ΔB caused by susceptibility changes $\Delta\chi$:

$$\frac{1}{T_2^*} \approx \frac{1}{T_2} + \gamma \cdot \Delta B = \frac{1}{T_2} + \gamma \cdot \Delta\chi \cdot B \qquad (5.1.1)$$

It should be noted that Eq. 5.1.1 assumes a model for the distribution of field offsets. In this sample the increase in field strength from 1.5 T to 3 T shortens the T_2*, but even at 3 T the T_2* values are longer than TE. Thus, the T_2* signal decay is still overcompensated by the SNR increase due to the higher field strength. For quantitative analysis, a high SNR is required for MRI of samples with short T_2*. Solenoid resonators used in this study are advantageous due to their high filling factor and sensitivity. However, very short T_2* values lower than 100 μs demand more advanced hardware modifications to further reduce acquisition delays. The relaxation maps indicate that for the ancient mummified sample used in this study, T_2* yields the differentiating information between different tissues rather than T_1.

The term T_2* is used to describe the combination of T_2 decay and line broadening by field inhomogeneities and susceptibility as shown in Eq. 5.1.1. Considering the typical field susceptibility differences, the dominant factor of the signal decay can be attributed to short T_2 rather than susceptibility and field inhomogeneity effects. Therefore, the mummified sample used in this study can be classified as a short T_2 sample instead of a short T_2*; however, no direct measure of T_2 was conducted since the acquired signals were FIDs rather than spin echoes.

UTE and PETRA images show that MRI of ancient mummified remains is feasible in clinical MR systems with a good tissue contrast. Clinical MRI systems are widely available and can be utilized, especially for mummy imaging, for overnight scans as mummies do not require special care during the scan. Nevertheless, acquisition times should be kept at a minimum to avoid systematic effects by B_0 field drift and field instabilities caused by slow gradient heating. Furthermore, some ancient samples should not be exposed to an excessive amount of humidity or UV light for prolonged periods of time, and therefore time-optimized protocols should be designed. In addition to MR applications for basic research on ancient mummies, its potential use would be for post mortem diagnostic imaging of modern, naturally dry mummified corpses which are described with an increasing frequency in the forensic literature [216]–[219].

Solenoid resonator designs are especially suitable for extremity imaging due to their high homogeneity and high filling factor. For whole body mummified specimen, however, solenoid resonators may not be optimal since they have to be oriented perpendicular to B_o. In that case, birdcage resonator designs would be more useful. To image smaller sub-regions, surface resonators could be used. In some image data weak signals from the resonator housings are detected even though they did not interfere with the analysis of the data. Proton-free resonator designs are needed to further reduce signals from the resonator housing.

Recently developed pulse sequences such as ZTE [220] and SWIFT [51] would also be of interest for MRI of mummified samples. These sequences can be implemented on pre-clinical hardware which provides short RF pulses, strong gradient amplitudes and very low dead-times on the order of a microsecond. Unfortunately, larger mummified samples cannot be imaged at preclinical MRI units. Clinical MRI systems, however, have dead-times of several tens of microseconds (here: 40 μs) which sets a lower limit to the switching between RF transmission and signal reception. Thus, the use of ZTE and SWIFT is currently impracticable at clinical MRI systems.

Gradient fidelities and limitations are critical for all short-TE sequences, in particular for SPI. In SPI, the minimum echo time is dependent on the gradient amplitude and the outermost point in k-space that is measured. With standard clinical hardware, this point is reached in 300 μs for the spatial resolution used in this study; novel gradient systems with G_{max} of 80 - 300 mT/m, however, might be able to reduce these encoding times significantly, and thus leading to reduced losses from T_2* decay [221]. Another limiting factor for both SPI and PETRA are potential distortions due to the amplitude modulation of the very short RF pulses (here: 10 μs): short pulses are difficult to realize with clinical MR hardware, and Tx resonator ring-down may additionally distort the rectangular pulse profile leading to longer pulses. Longer RF pulses are more susceptible to the presence of the encoding gradients, which means that the excitation becomes increasingly spatially selective. This may cause blurring of the MR signal which increases with distance from the magnet isocenter.

The sensitivity penalty of pure phase encode SPI and SPRITE measurements can be reduced with detection of multiple FID points per acquisition. Averaging of multiple points would give a factor of square root of number of averaged points increase in SNR in exchange for a decrease in resolution.

However, in this study, the single point approach is applied to highlight the difference between the pure phase and frequency encoding strategies.

The clinical utility of these sequences for example for bone imaging in combined PET/MR systems still needs to be evaluated. Radial sequences are prone to blurring artifacts due to signal decay during the image acquisition. SPI shows no blurring but due to its long scan durations it is not suitable for most clinical exams. UTE and PETRA have similar scan times, which are in the clinically acceptable range of less than 10 min as long as TR is kept short and low flip angles are used. Spatial localization in 3D is best performed with UTE, which is perfectly non-slice-selective, whereas PETRA and gradient-ramped SPI introduce some spatial selectivity due to the presence of a slice selection gradient during RF excitation. Although unwanted slice selection can be corrected to some extent by signal post-processing, a non-selective excitation profile in 3D sequences is preferable. Another aspect is that PETRA and SPI are very silent due to the use of continuous gradient-ramps, and they are thus preferred in situations where patient comfort is important.

5.1.2 Gradient Friendly Modifications in SPI

The proposed k-space ordering is more gradient-friendly in a way that strong discontinuities in the k-space are avoided leading to a reduction of gradient-induced eddy currents. Furthermore, the gradient coils carry less energy per unit time. Although temperature measurements do not represent actual gradient coil temperatures and many interfering parameters are ignored, the results clearly show that an optimal ordering of k-space can reduce gradient heating.

Lowering the gradient level during RF excitation improves the excitation profile while maintaining the low acoustic noise levels of the continuous gradient sequences since gradients are not completely switched off. Limiting gradient amplitudes can also be applied to projection reconstruction sequences such as ZTE or similarly PETRA. Here, low gradient amplitudes might result in lower number of missing center data points. Gradient friendly modifications in SPI sequence with gradient shape modification can also be applied to PETRA and ZTE sequences in general to reduce the slice selectivity.

5.2 Concurrent Excitation and Acquisition

5.2.1 Active Decoupling

Previously, hybrid coupler decoupling has been proposed by Idiyatullin and coworkers to reduce B_1-induced signal leakage in CEA MRI [53]. In their work they achieved a decoupling of 30 - 40 dB, which necessitated the use of additional subtraction techniques to be able to reconstruct meaningful MR images. In this work, higher decoupling was achieved; however, a subtraction was still needed to remove the remaining B_1-induced leakage signals. Nevertheless, the increased decoupling in this work provides a better use of the dynamic range of the receiver, which translates directly into an increased image dynamic range. Note that the dynamic range can be traded off against sweep rate in chirped acquisitions, and thus the dynamic range per unit time could be used as a figure of merit for CEA acquisitions.

In this work two subsequent data sets are acquired: One with the regular experimental setup and the other with the sample removed. Measuring the signal with an unloaded resonator is a commonly applied technique [10], although the leakage signal has also been described analytically in classical CW-NMR experiments [222]. This analytical approach would be advantageous for clinical applications since the additional measurements without resonator loading can be avoided.

Minimization of the transmission noise is an important factor during CEA experiments. To minimize RF transmit noise, the conventional RF power amplifiers with 400 mV_{pp} noise voltage were replaced with an LNA. The noise voltage of this LNA was less than 2 μV_{pp}; thus, for these experiments transmit noise was not the dominant factor in the acquired CEA signal. In general, a CEA system should always use very low-noise transmit systems, which might require replacing some components in the transmit chain of clinical MRI systems. The ultimate aim of providing transmit-receive isolation over 100 dB requires demanding hardware modifications as well. This includes employing amplifiers with flat gain over the excitation bandwidth, providing constant noise figures of less than 0.5 dB; additionally, the PLL clock multiplier phase noise of the D/A converter modules, phase noise and jitter of the reference oscillators of the synthesizers should be minimized [223]–[225] using high performance products of RF communication electronics such as the yttrium-iron-garnet oscillators [226] and the high electron mobility transistors [227]. In addition, advanced spurious signal control is required so that the overall noise floor of the signal generator unit is kept as low as possible. For imaging large bodies, higher transmit powers are required, which would lead to an increase in the noise floor, if these experiment would simply be up-scaled. However, using multiple small transmit resonators with LNAs, as done in this experiment, might yield sufficiently low noise floors – this will be investigated in future experiments.

In PA decoupling different RF transmit systems are used. These RF sources have uncorrelated noise characteristics, and it is therefore not possible to decouple them with PA decoupling alone. In this work an orthogonal resonator placement was used as an additional decoupling mechanism. Another concept would be to reintroduce noise correlation by using a single RF source and later split the RF power to both transmit resonators Tx coil-1 and 2. This would require a very stable RF amplifier, and additional phase and amplitude control units, but it would offer the opportunity to use the PA concept on every conventional MRI system without parallel RF transmitters.

To use the PA decoupling in clinical applications the PA decoupling must be stable over the total acquisition of the images. Decoupling instabilities could originate, for example, from transmit voltage fluctuations, frequency drifts in the modulators, phase noises based on oscillator infidelities, and time-varying cable interferences. These system instabilities can lead to time-varying leakage signals so that a reacquisition of the subtraction signal $A(t)$ might be required. This could be difficult to achieve in a clinical application, and thus it needs to be investigated in the future whether the original time course $A(t)$ can be adapted numerically (e.g., by scaling) to small changes in the decoupling. In these pilot experiments the acquisition time for a CEA data set was more than 2 min due to the relatively long repetition time of 200 ms, which was chosen to maximize the MR signal. In clinical routine much shorter TRs are used so that CEA MRI measurements will become less susceptible to long term fluctuations of the decoupling which are demonstrated in Fig. 62. If the B_1-induced leakage deviates at each acquired spoke, the resulting k-space would be contaminated by undefined frequency modulations which might results in artifacts effecting low or high frequency information content. Therefore, a more comprehensive treatment of B_1-induced leakage subtraction is necessary in advance to further applications of the method. For large-bandwidth applications (i.e., *in vivo*) it is important that the decoupling over the excitation bandwidth is sufficiently high to keep the signal within the dynamic range of the receiver.

During patient motion PA decoupling may be variable, as was reported for the change of resonator S-parameters and impedance due to breathing motion [228]–[230]. These motion-induced load changes were on the order of a few percent, which would not exceed the dynamic range of the receiver in our experiment. The difference in isolation between two extreme cases of a fully loaded resonator and an unloaded resonator was less than 20 dB, and the leakage signal was still within the receiver dynamic range. Nevertheless, a real-time monitoring of the loading (as was discussed in [231]) along with a feedback system would definitely increase the robustness of the active decoupling system against motion.

In this work custom-designed RF amplifiers were used. Even though care was taken to operate them in their linear regime, non-linearities cannot be fully excluded. In commercial MR systems RF non-linearities are compensated by calibration measurements and pre-emphasis, which is also possible with this implementation. Note, however, that small RF transmit non-linearities do not affect the deconvolution (and, thus, the image quality) as they are intrinsically accounted for by the measurement of B_1. For the PA decoupling, transmit nonlinearities might lead to a less optimal isolation in particular for RF pulses with a large dynamic range. In this case, decoupling would be dependent on the current RF pulse amplitude and phase, which varies over pulse.

In the reconstructed images blurring of the edges can be observed. Radial sampling and T_2* decay limits the full-width at half-maximum (FWHM) resolution [232]. For the rubber sample with a very short T_2* of 0.5 ms, a T_2* decay-dependent blurring of 8 pixels was expected. Additional blurring would be introduced due to the large effective slice thickness caused by 2D acquisition without slice selection, B_0 at the edges, and the angle between the holes axes and the imaging plane. If we assume a short T_2* of 10 ms (which is already shorter than typical T_2*s in tissue at 3T), the T_2* decay during data acquisition would lead to a widening of the point spread function FWHM by a factor of 0.2 – thus, the T_2* decay and 3D sampling cannot account for the observed blurring of more than 2 voxels. The blurring could be caused by small errors in the gradient trajectories which lead to misalignments of the k-space data. Another possible source of blurring could be a filtering effect in k-space: depending of the location in the imaged object, the actual time of the excitation within the sweep pulse varies, which leads to a spatially varying repetition time. This effect is small from one radial projection to the next, but it can introduce variable T_1 relaxation, and, thus, amplitude modulation. To improve image quality, gradient trajectories need to be mapped, and the k-space filtering effect could be removed by advanced pulse sequence strategies. Increased bandwidth is a direct way to decrease off-resonance blurring in radial imaging of fast relaxing spins, where, the trend of increasing the maximum gradient strength will help to improve CEA MRI.

In this work the position dependence of B_1 was not included in the analysis described by Equation 3. As the rubber sample was smaller than the Tx resonator, this B_1 inhomogeneity is not expected to influence the quality of the result significantly. Note further, that in the CEA experiment a frequency-swept RF pulse was used, which is similar to an adiabatic pulse for spin excitation [44]. Adiabatic pulses provide nearly homogeneous RF excitation even in the presence of inhomogeneous B_1, and thus the CEA experiment can be further improved by a careful selection of an adiabatic excitation pulse. With adiabatic pulses the small tip angle approximation is violated, but a reconstruction should still be possible, as the response of the spin system to the excitation pulse can be computed numerically.

The PA concept can be further expanded by the use of more than two Tx resonators that would potentially reduce the remaining B_1-induced voltage after PA decoupling. Unfortunately, these additional resonators can couple to the other Tx resonators which might reduce the decoupling efficiency. Numerical simulations or approximate analytical solutions are required to analyze this decoupling problem, which is beyond the scope of the present study. Nevertheless, the decoupling with two Tx coils is already good enough to provide CEA MR images with minimal B_1 leakage. Another alternative might be a direct Cartesian feedback of the Tx signal as proposed by Hoult et al. [233]. More advanced techniques including dynamic analog cancellation are also implemented throughout the thesis. PA decoupling can potentially be combined with other methods decoupling to achieve extra isolation.

At higher field strengths, PA decoupling can be implemented theoretically using the same approach presented here. However, at higher field strengths resonator and sample dimensions are closer to the RF wavelength, the coupling between Tx and Rx resonators may vary more strongly at small displacements of the sample (e.g., during breathing). Nevertheless, the decoupling is still a scalar, and the problem is conceptionally similar to B_1 inhomogeneity correction with parallel transmit systems. Here, the B_1 map of each Tx resonator is measured and the optimum phase/amplitude modulation parameters are calculated to minimize the induced voltage in the Rx resonator, whereas in B_1 inhomogeneity correction the B_1 field in the target area is made homogeneous. Thus, all existing technologies for B_1 homogenization can be adapted ranging from simple B_1 shimming to the individual modulation of the RF pulse shapes for each transmit channel to achieve maximum transmit-receive isolation.

As a further improvement it is possible to use an optimized RF pulse at Tx coil-2 so that the decoupling is maximized. Such a pulse could take into account additional effects such as system non-linearities, and it would bring more degrees of freedom to the setup.

The same setup and system can also be used for conventional pulsed MR sequences such as UTE. The concept of reducing the B_1-induced currents in UTE sequences was first presented as a potential solution to the acquisition delay problem in UTE sequences due to resonator ring-down time [234]. The acquisition delay after RF excitation is caused by T/R switching, ADC filtering, and the resonator ring-down time. Using the active decoupling method presented in this work, T/R switches can be eliminated from the setup because the method provides an isolation of 70 dB between Tx and Rx resonators with the initial setup of two Tx and one Rx resonators. In addition, this isolation decouples the Tx and Rx resonators during the excitation, resulting in a reduction in the energy that is deposited in the Rx resonator during excitation and a decrease in resonator ring-down time. Therefore, this decoupling method is a potential solution for the acquisition delay problem in UTE sequences.

Extension of the presented decoupling scheme to planar surface array instead of a single Rx loop resonator is challenging to implement. Initially one can start with a planar surface array and treat signal from each resonator element separately. Afterwards signal from all receive elements could be combined. Geometric decoupling is possible for all the resonator elements and a certain amount of decoupling could be achieved for each element although not optimal. It could also be the case that the system requires an at least equal number of transmitters and would still leave the question open, if the field produced by the transmitter array, while fully and actively decoupling it from the receiver, would still be efficient for spin excitation.

The extension to receive arrays is yet a problematic at this point of the research. Experimental validation is required to take a further step towards CEA MRI with receive arrays.

State-of-the-art gradients of clinical MRI systems offer high maximum gradient strengths (up to 80 mT/m) and high slew rates (typically up to 200 T/m/s). Combined with fast Tx/Rx switching circuits designed specifically to reduce acquisition delays [51], [220], high performance gradients can already achieve very broad bandwidths that cover extremely short T_2* components in vivo. However, rapid gradient switching is limited by physiology due to peripheral nerve stimulation [235], and increasing gradient strengths are more demanding in terms of system engineering. CEA, on the other hand, approaches the problem from a different perspective and reduces a few hardware demands such as the need for very fast switching circuits. In addition, CEA can evolve in line with parallel transmit array systems. Further studies are necessary to investigate the benefits of CEA in vivo and provide a sound comparison between state-of-the-art techniques for imaging of tissue with ultra-short T_2.

5.2.2 Automated Analog Cancellation using Dynamic Feedback from MRI System

5.2.2.1 Automated Feedback Decoupling System

The analog cancellation system proposed in this work provides a suppression of the leakage signal of about 62 dB, which is consistent with cancellation values found in full-duplex radio communication [148]. The combination of geometric decoupling [4] and the analog cancellation system always achieved a suppression of 89±2 dB, which was also used as a stopping criterion for the automatic feedback system. In clinical MRI systems this suppression is not sufficient, and thus additional digital subtraction during post-processing was necessary to achieve the required total isolation of 100 dB [142].

Both the analog cancellation circuit as well as the geometrically decoupled RF resonators show a frequency-dependent response. Therefore, the amount of decoupling is dependent on the frequency offset. To minimize this, the analog cancellation system can be modified to have a flatter frequency response; however, the frequency-dependence is mainly caused by the resonance behavior of the RF resonators, i.e. resonator tuning. For the resonators the frequency dependence of the isolation could be decreased using multistage broadband matching circuits [236].

Chirp RF pulses with sinusoidal modulated edges were used in our experiments. Due to the frequency dependent response of the decoupling system, the pulse amplitude was modeled in post processing to remove the residual transmit leakage from the acquired signal. When more complicated frequency modulations are applied e.g. Hyperbolic-Secant pulses, modeling of residual leakage would also be complicated. As long as the computational cost is affordable, an iterative regularization based approach to calculate the residual leakage by taking the frequency dependent modulations into account might be needed.

5.2.2.2 Transmit Noise

In CEA MRI, Tx noise is one of the most important factors as it is directly related to SNR. Maximum deviation of less than 10% from the receive noise floor was measured. Although this measurement suggests that analog cancellation circuit does not contribute to the system noise significantly, it requires further analysis to relate the system noise to the SNR. A first proof-of-principle circuit was designed without optimizing the components such as varactors and PIN diodes by choosing extremely low noise characteristics. The noise behavior of the system can be further improved.

Automated feedback was used with 8% duty cycle, which can be increased to achieve more iterations or perform faster. Since the analog cancellation system was well behaving, the number of iterations in gradient descent optimization was less than 100 for all the trial experiments. This is probably due to the fact that optimal starting points were assigned. This approach is simple and does not require a long look-up table that stores many possible phase and attenuation pairs as in [149]. The current system lacks feedback to the MR system i.e. there is no control over the sequence running environment. When the automated decoupling is triggered, the iterations are done within the running sequence, which may result in some of the spokes during decoupling iterations to be useless and extracted out of the reconstruction data space. Next step for a more practical implementation is to control the pulse sequence environment for acquisition of the missing spokes during decoupling iterations.

Another point that has never been addressed in CEA literature, to our knowledge, is cable-cable couplings. Especially in existence of unbalanced shield currents, cable couplings result in significant reduction in decoupling performance. When more than one Rx channels are used e.g. PUCs or Rx arrays, the receive circuitry and cables should be well separated both electrically and physically. In our setup for the PUC, a separate receiver plug was used.

5.2.2.3 Contrast Mechanisms

The phantom and wrist images suggest that there is a dominant proton density weighting along with a T_1 relaxation based contrast which can be deducted from the small $CuSO_4$ tube phantoms with different concentrations. The images compared to a high resolution GRE sequence. Even the proof-of-principle CEA image is promising for representing tendons with high signal intensity compared to GRE image where the tendons have signal void.

5.2.2.4 Receiver System

With increasing decoupling, the possibility of quantization errors in the received signal is decreased. Therefore, the task of the decoupling is not only to reduce the residual leakage down the Rx dynamic range, but also to make an efficient use of ADC dynamic range. Considering the frequency dependent nature of the residual leakage, ADC bits that define the 'clean' signal are always limited, thus the SNR. In order to overcome this problem, use of lock-in amplifiers is possible that involves phase sensitive detection. Lock-in amplifiers can be set to detect the convolved FID response alone, when an FM RF pulse is introduced as the reference signal. This way, it is possible to benefit from full ADC dynamic range for more efficient encoding of the MR signal.

In general, it is possible to use this system in combination with receive arrays. For receive array configuration, one possible setting is to use a separate analog cancellation circuit for each channel. However, in this configuration, couplings among the resonators and the cables will complicate the solution, and assigning a single phase and amplitude scale for each channel will not be as simple as the single resonator case. A more likely solution is designing an advanced version of the current automated feedback decoupling that is extended to the number of Rx elements. An intelligent algorithm can search for the optimal values for attenuation and phase.

5.2.2.5 Data Processing

Regularization based deconvolution performs better than simple Fourier transform methods in terms of noise and FID distortion in the examples shown in this study. However, the quality of the result strictly depends on the choice of the regularization function. A more comprehensive investigation is required to find the ideal reconstruction parameters. Moreover, instead of a projection based approach, image based iterative regularized reconstruction methods are more useful, as the ultimate aim is to obtain a high image quality. It is beyond scope of this thesis but as a future work, algebraic reconstruction techniques with iterative regularization methods needs to be implemented for automated reconstruction of the images. This will also enable examination of the experiment results right after the measurement.

5.2.2.6 Clinical Applications

CEA MRI with dynamic analog cancellation is an initial step to bring this method into clinical practice. First, it is time efficient due to the automated feedback decoupling operation; second, reduction in decoupling due to motion or changes in loading can be compensated in short time. Considering the advantages of 100% acquisition efficiency and extremely low peak RF powers, CEA MRI is a strong candidate to be a promising tool in various clinical applications including musculoskeletal imaging, connective tissue, and myelinated neurons.

In interleaved sequences, signals decaying more rapidly than the system dead times cannot be detected, whereas CEA continuously monitors the MR signal even during RF excitation. Thus, tightly bounded water in bone, for example, is not visible with interleaved NMR approaches [48], [51], [125], [135], [232]; yet new contrast mechanisms and novel sequences may be developed with further research on CEA.

5.3 EGG Measurements during Dynamic Lung MRI

The presented preliminary results suggest that EGG recording is possible during lung MRI, so that glottal activities can be correlated with EGG measurements. To assess the safety of the EGG device, four safety tests were performed. According to these currently established safety standards [237]–[241] the modified EGG could be classified as "MR Conditional", i.e. it poses no known hazards in a specified MR environment with specified conditions of use (cf. Methods, Volunteer Measurements). It should be noted that the safety standards are currently revised, and so this assessment will have to be repeated once new specifications are available.

In Fig. 70b, there is a more than linear temperature increase at later times over 15 min. Nevertheless, the temperature increase is 1.2 °C, which is considered acceptable. In the experiments with singers, duration of individual singing trial should not exceed 5 min, and under these conditions the temperature increase would be even less. One problem of the RF heating test is that the electrodes were immersed in a gel, whereas in a realistic scenario they are attached to the skin with one half of the electrodes being in contact with air. Unfortunately, a standard test method dedicated to the surface-skin contact electrodes in MRI is not available, and thus the applied test methods were adapted from those for medical implants. Local SAR distribution might differ for a more realistic scenario of surface-skin contact. However, real-time monitoring of temperature at the surface-skin contact showed no hazardous RF-induced heating ($\Delta T < 0.4°$ C), if the RF power was kept low ($P < 0.2$ W). This setting corresponded to a reduction in the SAR limit by a factor of 50, which can be challenging in some imaging situations. However, in our experiments this corresponded to a reduction of the true-FISP flip angle from 70° to 10°, which still provides MR images with very good contrast in the lung. In the future, results from RF simulations and new RF test methods might enable increasing this very conservative SAR limit.

With the repetition times used in dynamic MR imaging, the frequency of the RF-induced signal and the fundamental frequency of human voice are close to one another. Therefore it is not possible to use frequency filters to extract the undistorted EGG signal. For correction, in this proof-of-concept study ICA signal processing was used. ICA was feasible for a small amount of data but it required manual interpretation of the results; in the future, ICA post-processing needs to be automated so that it can handle all EGG data without for the requirement of a manual segmentation and subjective interpretation. ICA was found to be a robust way to separate the two statistically independent data, the RF induced signal and the EGG signal. Using a dual-channel EGG device this method could be applied more effectively, where two electrode pairs acquire the EGG signal and the RF induced signal envelope as different superpositions. One can even leave the second electrode pair floating in order to acquire only the RF induced envelope. ICA could be directly applied on the acquired 2 channel data to extract the corrected EGG signal automatically.

The subject started singing at a high pitch where more subglottic pressure was needed [165], then a two-octaves jump necessitated reduction of subglottic pressure for singing a lower pitch which resulted in constriction of the diaphragm. It could be speculated, that the diaphragm was used to suddenly reduce the subglottic pressure. Additionally, as expected, singing a high pitch causes more rapid lung volume reduction than low pitch. This data suggest that MRI in combination with EGG is a useful tool in the evaluation of breathing strategies in singers. Combination of EGG information and dynamic behavior of lungs and diaphragm is expected to help researchers to have a better understanding of singing physiology.

Appendix

A.1 Construction of BALUNs

In this section, construction of three different sorts of BALUNs will be explained.

Solenoid BALUN

An inductor of few windings using a coaxial cable is formed as in Fig. A1. A semi-rigid coaxial can be used, or alternatively, the coaxial cable connected to the resonator itself can be wound to form an inductance. In parallel to the formed inductance, a capacitor is soldered on the shield or on a PCB as in Fig. A1. After tuning the resonant circuit at the required frequency e.g. Larmor frequency, the whole structure is covered using a conducting cage.

Fig. A.1: Solenoid BALUN part of an active detuning circuit. A resonant LC circuit is formed on the ground line to suppress unbalanced shield currents.

A practical method of tuning the BALUN is to combine the grounds at the open ends of two coaxial cables, leaving the inner conductors opened and separate as shown in Fig. A2. The cables are connected to two ports of the network analyzer, and the device is set for S21 measurement. The tips of the inner conductors are touched on two poles of the tuning capacitor, and minimum S21 at the required frequency is searched (Fig. A2).

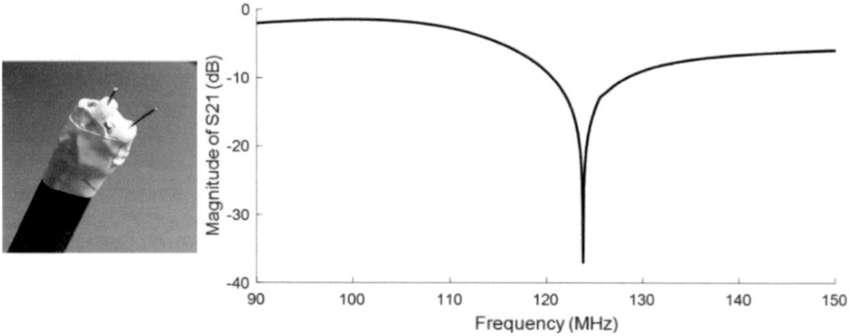

Fig. A.2: Measurement of a BALUN using an S21 common-ground probe (left) touched to the capacitor poles, and an exemplary S21 data acquired using a network analyser (right).

Bazooka BALUN

A convenient way to implement this BALUN is using tri-axial cable. However, due to its size and cost tri-axial cable is not always used in RF resonators. Alternatively a short cylindrical mechanical former can be constructed to simulate the tri-axial line. The blocking impedance is almost proportional to the length of the former and $log(R/r)$, where R and r are the radii of the former and the cable, respectively. Outermost conductor of the tri-axial cable or BALUN former is connected to the outer conductor of the cable. At distance of quarter wavelength another connection is made, or alternatively, a capacitor is connected between two conductors at a shorter distance than quarter wavelength. Tuning is done in the same way as the solenoid BALUN. The disadvantage of this BALUN is its low blocking impedance. A sample bazooka BALUN formed on a tri-axial cable and corresponding S21 plot is shown in Fig. A3.

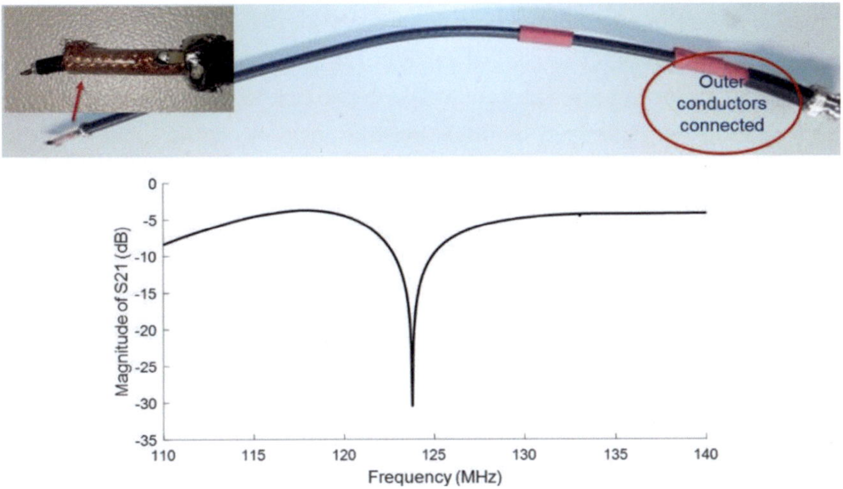

Fig. A.3: Bazooka BALUN was formed using a tri-axial cable (top), and corresponding S21 plot (bottom).

Floating BALUN

One major advantage of the floating BALUN is that it can be added to the existing resonator setups without soldering. A cylindrical plastic pipe is divided into two pieces along the longitudinal axis. Each piece is covered by a copper tape (Fig. A4). In [205], floating BALUNs were described in detail.

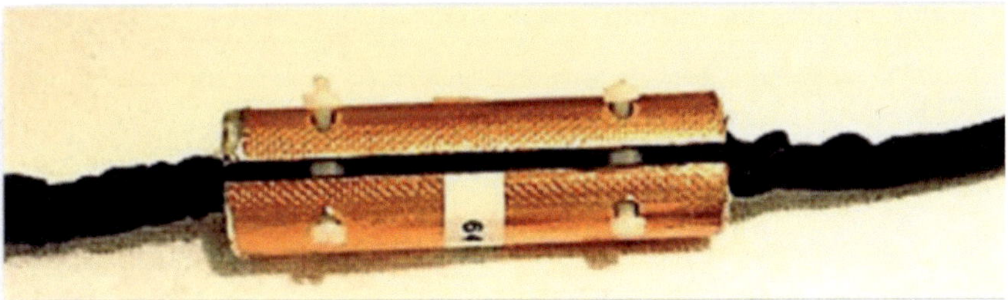

Fig. A.4: Floating BALUN placed on a shielded cable.

To determine the best position of the floating BALUN, two pick-up loops are placed or put through the coaxial cable at different positions and the floating BALUN is placed between two measuring points. S21 measurement using pick-up loops is minimized by adjusting the position on the cable and the screws of the BALUN.

B.1 List of Tissues with Short T_2

This list is prepared using the data compiled from the references [97], [117], [125], [242].

Tissue	T_2 (μs)
Ligaments	4000-10000
Achilles tendon	4000-7000
Knee menisci	5000-8000
Periosteum	5000-11000
Cortical bone	400-500
Dentine	150
Dental enamel	70
Protons in water tightly bound to proteins	10
Protons in proteins	10
Protons in solids e.g. calcium hydroxyapatite	≤ 1

Tab. A.1: T_2 values. The sources are adult clinical results and tissue sample results estimated for 1.5 T.

C.1 Gradient Delay Calculation Script

This Matlab function calculates the gradient delay in µs given two spokes s1, s2, which correspond to two opposite radial spoke acquisition in k-space. First, Fourier transforms are calculated. Then, cross-correlation function is calculated using conjugate multiplication. Derivative of the phase of the cross-correlation function is used to estimate the corresponding shift between two spokes.

```
function shift=grad_shift_calc(s1,s2)

    s1m=abs(s1);s2flipm=abs(fliplr(s2));
    S1=fftshift(fft(fftshift(s1m)));
    S2=fftshift(fft(fftshift(s2flipm)));
    gsp12=S1.*conj(S2);
    gsp12_ph=angle(gsp12);
    k=diff(unwrap(gsp12_ph));
    figure;plot(abs(gsp12));hold
on;plot(unwrap(gsp12_ph));plot(k);hold off;
    title('Gradient-delay correction'); legend('|g|','g-
ph','k');
    xlabel('Data samples');ylabel('Amplitude (a.u.)');
    Support=abs(gsp12);
    idx = Support>0.2*max(Support);
    ind_fit=find(idx);
    ph_lin=gsp12_ph(ind_fit);
    [ph_fit]=polyfit(ind_fit,ph_lin,1);
    ph_slope=ph_fit(1);
    shift=-1*ph_slope*size(s1,2)/(4*pi);
end
```

D.1 Quadrature Tx/Rx Switch

Advanced Design Systems (2015.01, Keysight Technologies, Santa Rosa, CA) circuit design tool was used to estimate lumped element components of the hybrid coupler (Fig. A5).

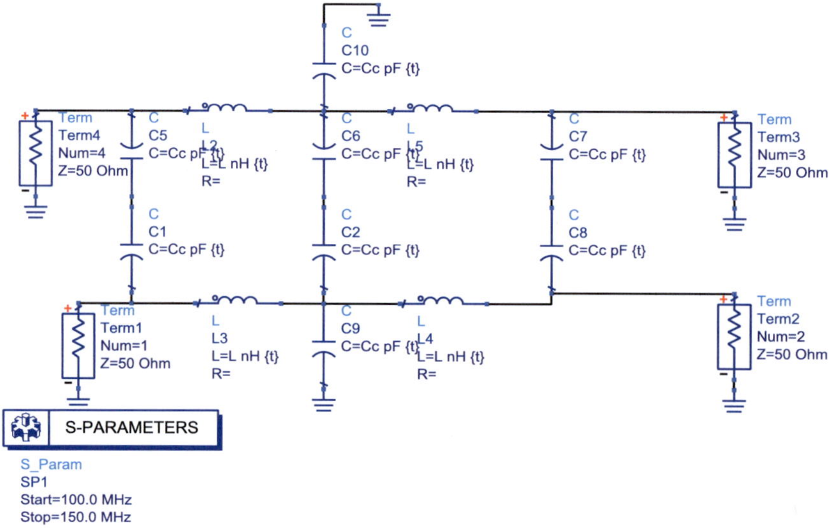

Fig. A.5: Hybrid coupler unit of the quadrature Tx/Rx switch. Capacitance of 25.9 pF and inductance of 65 nH were estimated for optimal performance at 12 MHz.

S-parameter simulations are shown Fig.A6.

Circuit schematic and S parameter simulations for the whole circuit is shown in Fig. A7 and Fig. A8.

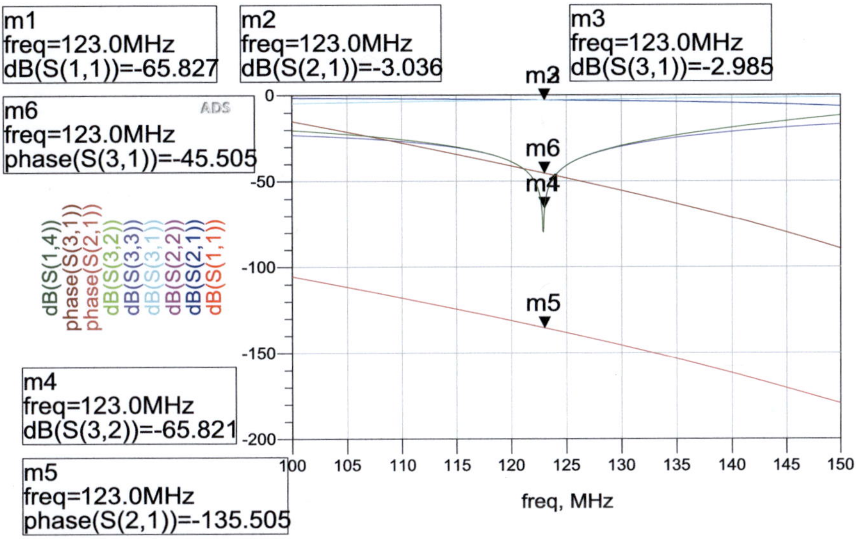

Fig. A.6: S parameter simulations for the hybrid coupler. Input power is equally divided between the ports 2 and 3, however with a 90° phase difference. Reflection coefficients and return losses are low enough to avoid unwanted signal leaks between transmit and receive ports.

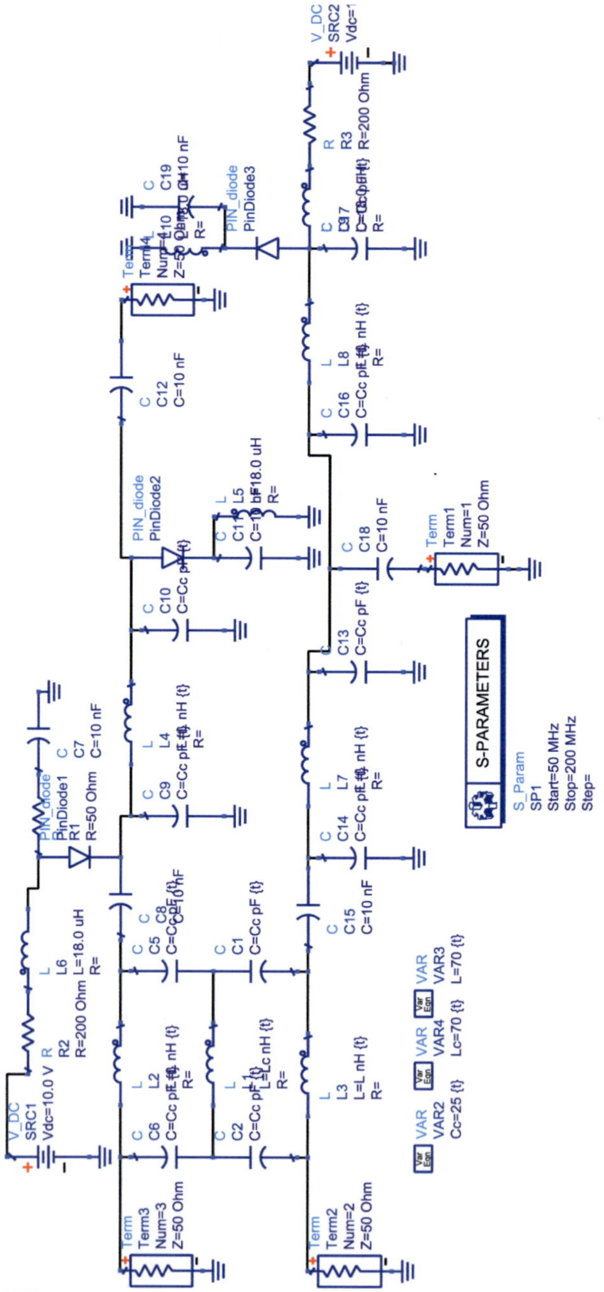

Fig. A.7: Quadrature Tx/Rx circuit diagram.

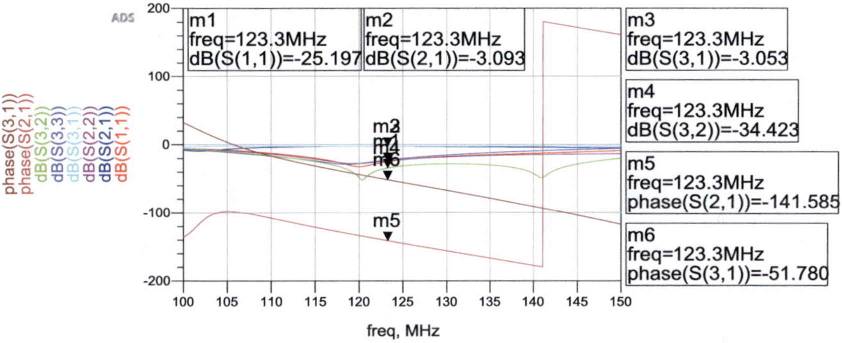

Fig. A.8: Quadrature Tx/Rx switch S-parameter simulation results. Isolation between Tx and Rx ports is 34.4 dB in Rx mode at 123.3 MHz (i.e. PIN diodes are in OFF state).

PCB layout for the whole circuit, as well as the photo of the prototype is shown in Fig. A9.

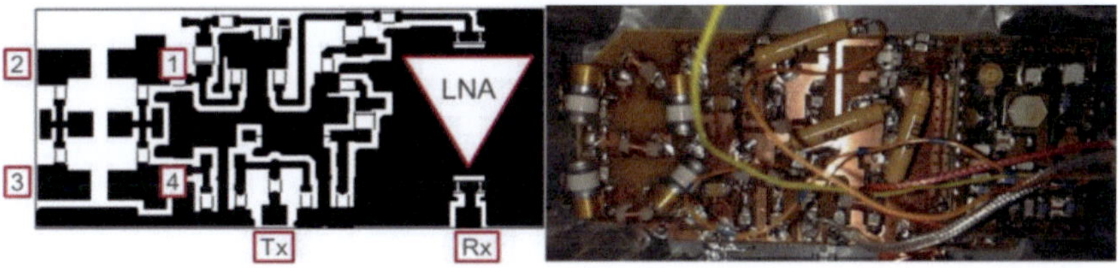

Fig. A.9: PCB Layout of 50x112 mm^2 (left), and photo of the prototype circuit for the quadrature Tx/Rx switch. Numbers 1-4 shows the hybrid coupler ports.

Test bench measurements using network analyzer verified the Tx/Rx switch function both in Tx and Rx mode (Fig. A10).

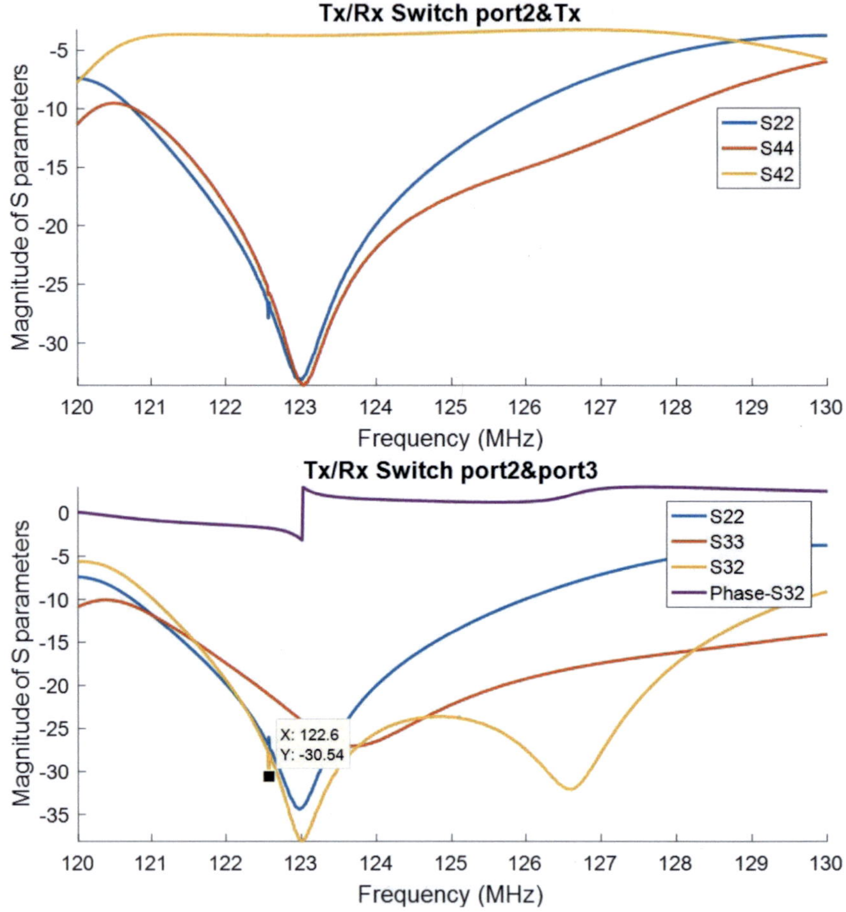

Fig. A.10: S parameter measurements between port 2 and Tx port (top), and port 3 and port 2 (bottom) of the quadrature Tx/Rx switch. S42 measurements verify that half of the power is transferred to the second port. S32 plots suggest that an isolation of more than 25 dB is secured between two ports connected to the resonator. Phase difference between S42 and S43 measurements is calculated as 87.7° at 123 MHz.

D.2 Linear Tx/Rx Switch

PCB layout and photo of the prototype of the linear Tx/Rx switch are given in Fig. A11.

Fig. A.11: PCB layout of 70x90 mm^2 (left) and photo of the prototype of the linear Tx/Rx switch (right).

E.1 Double-Angle Method Matlab Script

This script uses dicom image data as input. First, a corner region where no MR signal is present is selected. Then, standard deviation of the pixels in this 'noise' region is calculated. The pixel SNR is equal to the pixel signal intensity divided by the standard deviation of the noise region. A manually selected threshold nulls the pixels with lower SNR. The masked image data are used to calculate the flip angle, then finally, B_1 is estimated using the RF pulse shape parameters.

```
% masking:
noiSEct=I1((length(I1)-
round(length(I1)/5)):end,1:round(length(I1)/5));
[a, b]=size(noiSEct);
nstd=std(double(reshape(noiSEct,a*b,1)));
Isnr=(I1/nstd);
I1(Isnr<max(max(Isnr))/15)=0;
I2(Isnr<max(max(Isnr))/15)=0;
% slice=11;
% I=squeeze(I1(slice,:,:));
I=I1;
% Is2=squeeze(I2(slice,:,:));
Is2=I2;
%                                   noiSEct=I((length(I)-
round(length(I)/10)):end,1:round(length(I)/10));
%   [a, b]=size(noiSEct);
%   nstd=std(double(reshape(noiSEct,a*b,1)));
%   maskSNR=(double(I/nstd)>15);
% %   figure(62);imagesc(maskSNR);

%Form the data sets
r=double(I)./double(Is2);
IB1=(1/(42.756e6*1000e-6))*acos(1./(2*r)).*maskSNR;
IFA=acos(0.5./r).*maskSNR;
% IB1(Isnr<max(max(Isnr))/15)=0;
figure(61);imagesc(abs(IFA));title('Flip Angle map for
the CEA resonator 3 T');colormap(hot(256));axis im-
age;colorbar;

%                                                     fig-
ure;imagesc(abs(squeeze(1e3*IB1(48,:,:))));title('B1 map
for new solenoid resonator');colormap(hot(256));axis im-
age;colorbar;
% figure;imagesc(1e3*abs(IB1));title('B1 map for new so-
lenoid      resonator');colormap(hot(256));axis       im-
age;colorbar;
```

F.1 LNA Design

Circuit diagram for a single stage LNA design using a BFG541 bipolar transistor is shown in Fig. A12. Corresponding simulation results are shown in Fig. A13.

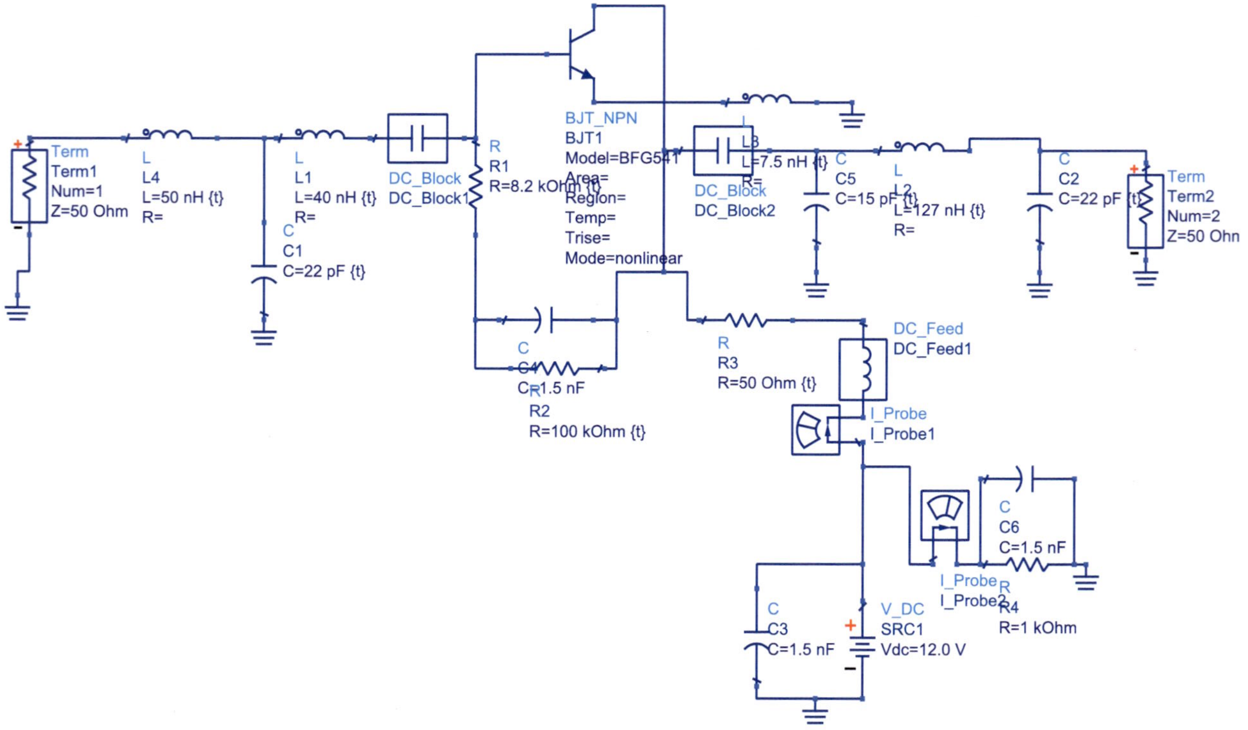

Fig. A.12: Circuit diagram of the LNA. Emitter inductor is used to stabilize the LNA. Bias resistors determine the gain. For input and output matching, LC tanks of three components are used.

Noise figure measurements are done using Agilent EXA N9010A Signal Analyzer (Agilent Technologies, Inc., USA). When supplied with 4 V and 8 V DC voltages, 1.6 dB and 1.8 dB NF were measured, respectively. Output linearity is maintained up to input voltage level of -8.93 dBm at 8V DC supply.

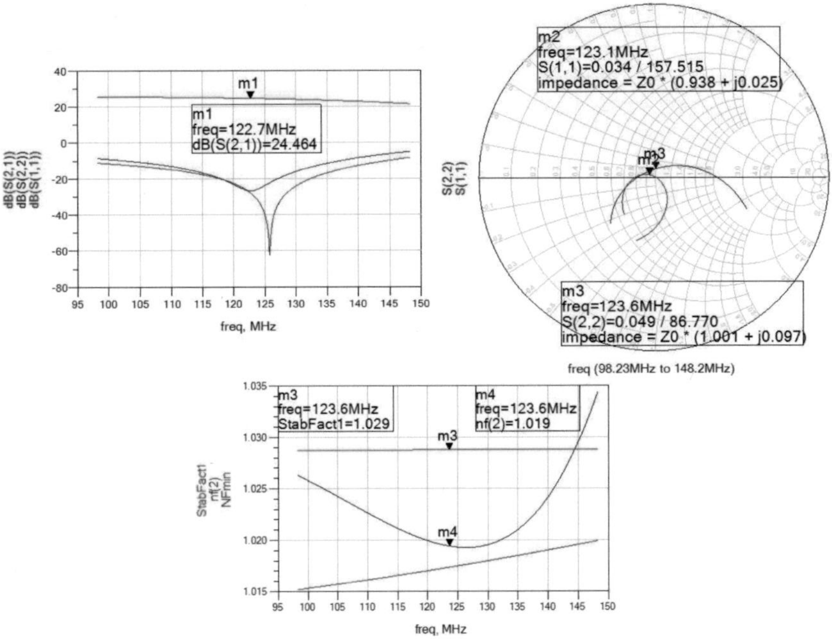

Fig. A.13: LNA simulation results. S parameter magnitude plots (left) shows a gain of 24.46 dB. Input and output matching conditions are shown in smith chart plot (middle). Stability factor and noise figure plots shows that the LNA is unconditionally stable and the noise figure is less than 1.02 dB between 120 MHz and 132 MHz.

PCB layout and photo of the prototype of the LNA circuit is shown in Fig. A14.

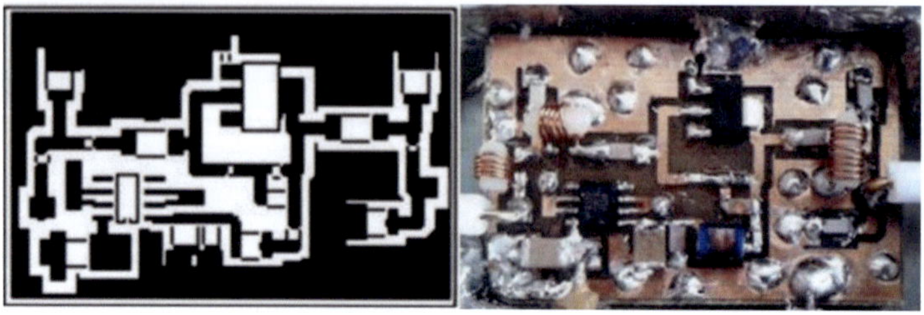

Fig. A.14: PCB Layout of 44x30 mm^2 (left), and the photo of the prototype LNA circuit (right).

Network analyzer measurements are shown in Fig. A15.

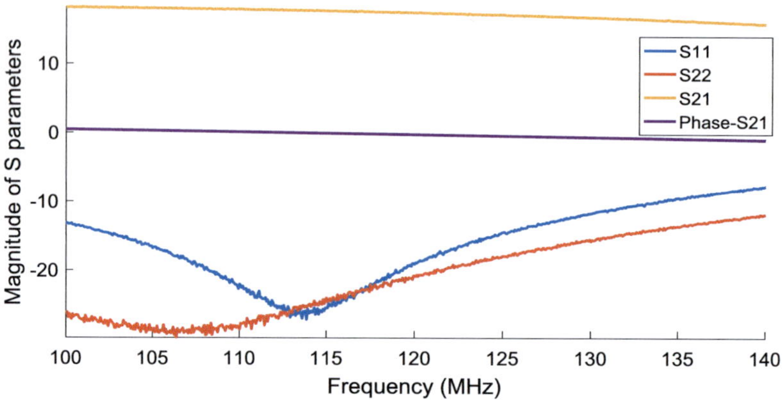

Fig. A.15: 17.4 dB gain at 123 MHz with 0.47 dB gain flatness within 10 MHz band is measured.

G.1 Analog Cancellation Circuit Design

Circuit diagrams for the two different types of analog cancellation circuits are shown in Fig. A16a and Fig. A16b.

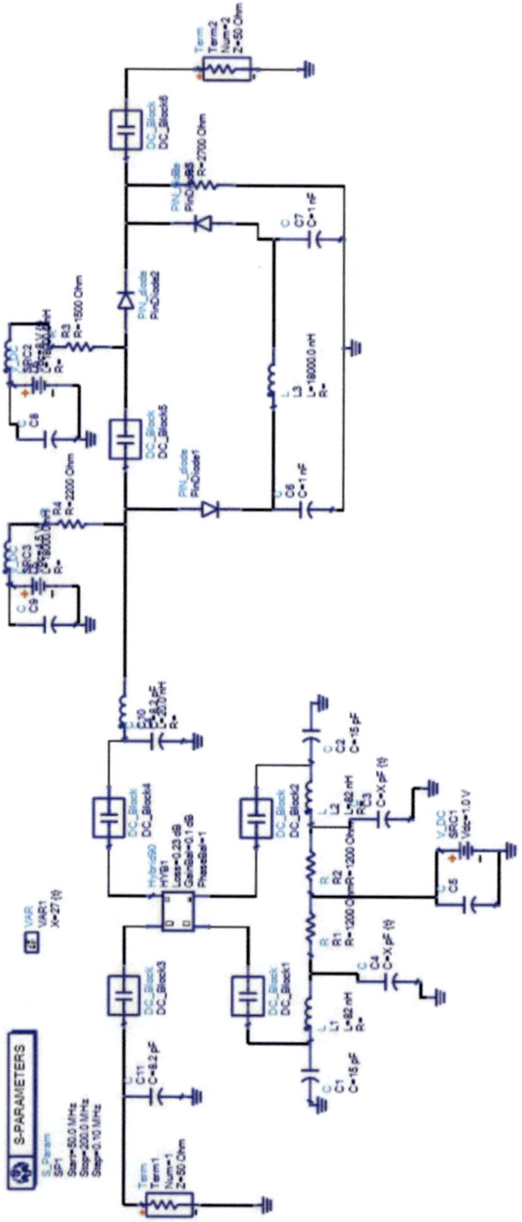

Fig. A.16a: Analog cancellation circuit diagram with PIN diode based attenuator.

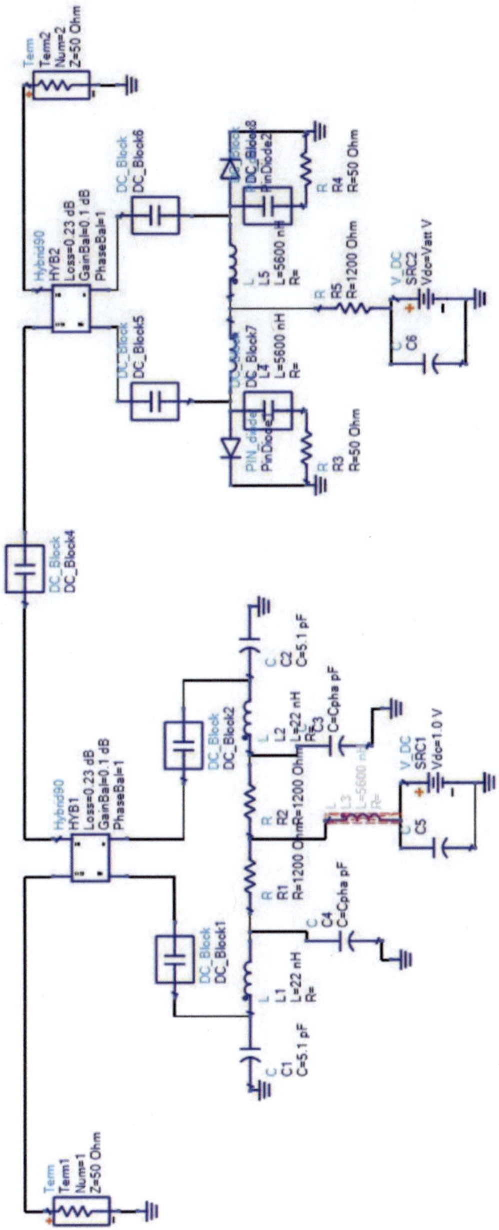

Fig. A.16b: Analog cancellation circuit diagram with hybrid coupler based self-matched attenuator.

PCB layouts and photos of the prototype circuits are shown in Fig. A17 for the self-matched design.

Fig. A.17: PCB layout of 118x60 mm^2 (left) and the photo (right) of the analog cancellation circuit with PIN-diode based attenuator.

H.1 Reconstruction Scripts for UTE and CEA Data

Reconstruction of UTE and CEA data are performed in a similar way, once the CEA data is pre-processed and projection data is obtained. Therefore, reconstruction of both is merged in a single main script where UTE or CEA option is selected in advance. After pre-processing, the remaining operation is re-gridding of the radial data on to a 3D Cartesian space. Here, one difference between UTE and CEA is the ramp-up sampling, which is performed during UTE acquisition, whereas in CEA, data acquisition starts during flat gradient strength. This is however handled automatically when CEA option is selected. Detailed explanation is given inside the scripts and functions which are available in https://github.com/alibaz/ConExAc.

Bibliography

[1] E. Fukushima and S. B. W. Roeder, *Experimental pulse NMR: a nuts and bolts approach*. Addison-Wesley Pub. Co., Advanced Book Program, 1981.

[2] I. I. Rabi, J. R. Zacharias, S. Millman, and P. Kusch, "A New Method of Measuring Nuclear Magnetic Moment," *Phys. Rev.*, vol. 53, no. 4, pp. 318–318, Feb. 1938.

[3] E. M. Purcell, H. C. Torrey, and R. V. Pound, "Resonance Absorption by Nuclear Magnetic Moments in a Solid," *Phys. Rev.*, vol. 69, no. 1–2, pp. 37–38, Jan. 1946.

[4] F. Bloch, W. W. Hansen, and M. Packard, "The Nuclear Induction Experiment," *Phys. Rev.*, vol. 70, no. 7–8, pp. 474–485, Oct. 1946.

[5] E. L. Hahn, "Spin Echoes," *Phys. Rev.*, vol. 80, no. 4, pp. 580–594, Nov. 1950.

[6] P. C. Lauterbur, "Image formation by induced local interactions. Examples employing nuclear magnetic resonance.," *Nature*, vol. 242, pp. 190–191, Mar. 1973.

[7] B. Blümich, P. Blünder, C. Fülber, U. Scheler, and F. Weigand, "NMR Imaging and materials research," *Macromol. Symp.*, vol. 87, no. 1, pp. 187–193, Nov. 1994.

[8] J. Dadok and R. F. Sprecher, "Correlation NNW spectroscopy," *J. Magn. Reson.*, vol. 13, no. 2, pp. 243–248, Feb. 1974.

[9] R. K. Gupta, J. A. Ferretti, and E. D. Becker, "Rapid scan Fourier transform NMR spectroscopy," *J. Magn. Reson.*, vol. 13, no. 3, pp. 275–290, Mar. 1974.

[10] D. J. Lurie, S. J. Mccallum, J. M. S. Hutchison, and M. Aleccit, "Continuous-wave NMR imaging of solids," *MAGMA Magn. Reson. Mater. Physics, Biol. Med.*, vol. 4, no. 1, pp. 77–81, Mar. 1996.

[11] G. R. Davies, D. J. Lurie, J. M. S. Hutchison, S. J. McCallum, and I. Nicholson, "Continuous-Wave Magnetic Resonance Imaging of Short T2 Materials," *J. Magn. Reson.*, vol. 148, no. 2, pp. 289–297, Feb. 2001.

[12] A. J. Fagan, G. R. Davies, J. M. S. Hutchison, and D. J. Lurie, "Continuous wave MRI of heterogeneous materials," *J. Magn. Reson.*, vol. 163, no. 2, pp. 318–324, Aug. 2003.

[13] A. J. Fagan, G. R. Davies, J. M. S. Hutchison, F. P. Glasser, and D. J. Lurie, "Development of a 3-D, multi-nuclear continuous wave NMR imaging system," *J. Magn. Reson.*, vol. 176, no. 2, pp. 140–150, Oct. 2005.

[14] A. Kumar, D. Welti, and R. R. Ernst, "NMR Fourier zeugmatography," *J. Magn. Reson.*, vol. 18, no. 1, pp. 69–83, Apr. 1975.

[15] OECD, "Magnetic resonance imaging (MRI) exams," 2016. [Online]. Available: http://www.oecd-ilibrary.org/social-issues-migration-health/magnetic-resonance-imaging-mri-exams/indicator/english_1d89353f-en. [Accessed: 03-Jun-2016].

[16] W. R. Nitz and P. Reimer, "Contrast mechanisms in MR imaging.," *Eur. Radiol.*, vol. 9, no. 6, pp. 1032–46, 1999.

[17] J. H. Duyn and A. P. Koretsky, "Novel frontiers in ultra-structural and molecular MRI of the brain.," *Curr. Opin. Neurol.*, vol. 24, no. 4, pp. 386–93, Aug. 2011.

[18] M. Uecker, S. Zhang, D. Voit, K.-D. Merboldt, and J. Frahm, "Real-time MRI: recent advances using radial FLASH," *Imaging Med.*, vol. 4, no. 4, pp. 461–476, Aug. 2012.

[19] Y. Eryaman, B. Guerin, C. Akgun, J. L. Herraiz, A. Martin, A. Torrado-Carvajal, N. Malpica, J. a Hernandez-Tamames, E. Schiavi, E. Adalsteinsson, and L. L. Wald, "Parallel transmit pulse design for patients with deep brain stimulation implants," *Magn. Reson. Med.*, vol. 73, no. 5, pp. 1896–1903, May 2015.

[20] S. Ogawa, T. M. Lee, A. R. Kay, and D. W. Tank, "Brain magnetic resonance imaging with contrast dependent on blood oxygenation.," *Proc. Natl. Acad. Sci.*, vol. 87, no. 24, pp. 9868–9872, Dec. 1990.

[21] L. Lord, P. Expert, J. F. Huckins, and F. E. Turkheimer, "Cerebral energy metabolism and the brain's functional network architecture: an integrative review," *J. Cereb. Blood Flow Metab.*, vol. 33, no. 9, pp. 1347–1354, Sep. 2013.

[22] K. P. Pruessmann, M. Weiger, M. B. Scheidegger, and P. Boesiger, "SENSE: sensitivity encoding for fast MRI.," *Magn. Reson. Med.*, vol. 42, no. 5, pp. 952–62, Nov. 1999.

[23] J. Hennig, A. Nauerth, and H. Friedburg, "RARE imaging: A fast imaging method for clinical MR," *Magn. Reson. Med.*, vol. 3, no. 6, pp. 823–833, Dec. 1986.

[24] J. Hennig and K. Scheffler, "Hyperechoes," *Magn. Reson. Med.*, vol. 46, no. 1, pp. 6–12, Jul. 2001.

[25] M. Murphy, M. Alley, J. Demmel, K. Keutzer, S. Vasanawala, and M. Lustig, "Fast L1-SPIRiT Compressed Sensing Parallel Imaging MRI: Scalable Parallel Implementation and Clinically Feasible Runtime," *IEEE Trans. Med. Imaging*, vol. 31, no. 6, pp. 1250–1262, Jun. 2012.

[26] P. A. Bottomley, W. A. Edelstein, W. M. Leue, H. R. Hart, J. F. Schenck, and R. W. Redington, "Head and body imaging by hydrogen nuclear magnetic resonance," *Magn. Reson. Imaging*, vol. 1, no. 2, pp. 69–74, Jan. 1982.

[27] C. J. Bergin, J. M. Pauly, and A. Macovski, "Lung parenchyma: projection reconstruction MR imaging.," *Radiology*, vol. 179, no. 3, pp. 777–81, Jun. 1991.

[28] S. Hafner, "Fast imaging in liquids and solids with the Back-projection Low Angle ShoT (BLAST) technique," *Magn. Reson. Imaging*, vol. 12, no. 7, pp. 1047–1051, Jan. 1994.

[29] D. P. Madio and I. J. Lowe, "Ultra-fast imaging using low flip angles and fids," *Magn. Reson. Med.*, vol. 34, no. 4, pp. 525–529, Oct. 1995.

[30] M. F. Dempsey and B. Condon, "Thermal Injuries Associated with MRI," *Clin. Radiol.*, vol. 56, no. 6, pp. 457–465, Jun. 2001.

[31] D. I. Hoult and P. C. Lauterbur, "The sensitivity of the zeugmatographic experiment involving human samples," *J. Magn. Reson.*, vol. 34, no. 2, pp. 425–433, May 1979.

[32] F. Bloch, "Nuclear Induction," *Phys. Rev.*, vol. 70, no. 7–8, pp. 460–474, Oct. 1946.

[33] D. Nishimura, *Principles of magnetic resonance imaging*. Stanford (CA): Stanford University Press, 1996.

[34] A. N. Garroway, P. K. Grannell, and P. Mansfield, "Image formation in NMR by a selective irradiative process," *J. Phys. C Solid State Phys.*, vol. 7, no. 24, pp. L457–L462, Dec. 1974.

[35] P. Hasgall, F. Di Gennaro, C. Baumgartner, E. Neufeld, M. Gosselin, D. Payne, A. Klingenböck, and N. Kuster, "IT'IS Database for thermal and electromagnetic parameters of biological tissues, Version 3.0," 2015. [Online]. Available: www.itis.ethz.ch/database. [Accessed: 01-May-2016].

[36] B. Cowan, *Nuclear Magnetic Resonance and Relaxation*. Cambridge: Cambridge University Press, 2005.

[37] D. Shaw and NMR, *Fourier Transform N.M.R. Spectroscopy*. Amsterdam, New York: Elsevier, 1984.

[38] G. A. Morris and P. B. Chilvers, “General Analytical Solutions of the Bloch Equations,” *J. Magn. Reson. Ser. A*, vol. 107, no. 2, pp. 236–238, Apr. 1994.

[39] P. T. Callaghan, *Principles of Nuclear Magnetic Resonance Microscopy*. Oxford: Clarendon Press, 1993.

[40] D. I. Hoult and R. E. Richards, “The signal-to-noise ratio of the nuclear magnetic resonance experiment,” *J. Magn. Reson.*, vol. 24, no. 1, pp. 71–85, Oct. 1976.

[41] R. R. Ernst and W. A. Anderson, “Application of Fourier Transform Spectroscopy to Magnetic Resonance,” *Rev. Sci. Instrum.*, vol. 37, no. 1, pp. 93–102, 1966.

[42] E. L. Hahn, “Spin Echoes,” *Phys. Rev.*, vol. 80, no. 4, pp. 580–594, Nov. 1950.

[43] A. Haase, J. Frahm, D. Matthaei, W. Hänicke, and K.-D. Merboldt, “FLASH imaging: Rapid NMR imaging using low flip-angle pulses,” *J. Magn. Reson.*, vol. 213, no. 2, pp. 533–541, Dec. 2011.

[44] A. Tannús and M. Garwood, “Adiabatic pulses.,” *NMR Biomed.*, vol. 10, no. 8, pp. 423–34, Dec. 1997.

[45] A. Tannús and M. Garwood, “Improved Performance of Frequency-Swept Pulses Using Offset-Independent Adiabaticity,” *J. Magn. Reson. Ser. A*, vol. 120, no. 1, pp. 133–137, May 1996.

[46] J. M. Bohlen and G. Bodenhausen, “Experimental Aspects of Chirp NMR Spectroscopy,” *J. Magn. Reson. Ser. A*, vol. 102, no. 3, pp. 293–301, May 1993.

[47] M. S. Silver, R. I. Joseph, and D. I. Hoult, “Highly selective and π pulse generation,” *J. Magn. Reson.*, vol. 59, no. 2, pp. 347–351, Sep. 1984.

[48] M. Garwood, “MRI of fast-relaxing spins,” *J. Magn. Reson.*, vol. 229, no. 0, pp. 49–54, Apr. 2013.

[49] M. Garwood and L. DelaBarre, “The Return of the Frequency Sweep: Designing Adiabatic Pulses for Contemporary NMR,” *J. Magn. Reson.*, vol. 153, no. 2, pp. 155–177, Dec. 2001.

[50] C. P. Poole, *Electron Spin Resonance*, 2nd ed. New York, USA: Wiley, 1983.

[51] D. Idiyatullin, C. Corum, J.-Y. Park, and M. Garwood, “Fast and quiet MRI using a swept radiofrequency,” *J. Magn. Reson.*, vol. 181, no. 2, pp. 342–349, Aug. 2006.

[52] A. C. Özen, N. K. Ertan, and E. Atalar, "Detection of MR Signal during RF Excitation using a Transmit Array System," in *Proceedings of the 20th Scientific Meeting, International Society of Magnetic Resonance in Medicine*, 2012, p. 2295.

[53] D. Idiyatullin, S. Suddarth, C. a Corum, G. Adriany, and M. Garwood, "Continuous SWIFT," *J. Magn. Reson.*, vol. 220, no. April, pp. 26–31, Jul. 2012.

[54] O. Brunner, B. E. Dietrich, M. Pavan, and K. Pruessmann, "MRI with Sideband Excitation: Application to Continuous SWIFT," in *Proceedings of the 20th Scientific Meeting, International Society of Magnetic Resonance in Medicine*, 2012, p. 150.

[55] F. Bloch and A. Siegert, "Magnetic Resonance for Nonrotating Fields," *Phys. Rev.*, vol. 57, no. 6, pp. 522–527, Mar. 1940.

[56] E. Saff and K. ABJ, "Distributing Many Points on a Sphere," *Math Intell*, vol. 19, no. 1, pp. 5–11, 1997.

[57] J. D. O'Sullivan, "A Fast Sinc Function Gridding Algorithm for Fourier Inversion in Computer Tomography," *IEEE Trans. Med. Imaging*, vol. 4, no. 4, pp. 200–207, Dec. 1985.

[58] D. C. Noll, J. M. Pauly, C. H. Meyer, D. G. Nishimura, and a Macovski, "Deblurring for non-2D Fourier transform magnetic resonance imaging.," *Magn. Reson. Med.*, vol. 25, no. 2, pp. 319–333, 1992.

[59] J. Murphy-Boesch and A. P. Koretsky, "An in vivo NMR probe circuit for improved sensitivity," *J. Magn. Reson.*, vol. 54, no. 3, pp. 526–532, Oct. 1983.

[60] E. R. Andrew and K. Jurga, "NMR probe with short recovery time," *J. Magn. Reson.*, vol. 73, no. 2, pp. 268–276, Jun. 1987.

[61] I. J. Lowe and C. E. Tarr, "A fast recovery probe and receiver for pulsed nuclear magnetic resonance spectroscopy.," *J. Sci. Instrum.*, vol. 1, no. 3, pp. 320–322, Mar. 1968.

[62] H. Nyquist, "Thermal Agitation of Electric Charge in Conductors," *Phys. Rev.*, vol. 32, no. 1, pp. 110–113, Jul. 1928.

[63] T. W. Redpath and C. J. Wiggins, "Estimating achievable signal-to-noise ratios of MRI transmit-receive coils from radiofrequency power measurements: applications in quality control," *Phys. Med. Biol.*, vol. 45, no. 1, pp. 217–227, Jan. 2000.

[64] D. I. Hoult, "The principle of reciprocity in signal strength calculations-A mathematical guide," *Concepts Magn. Reson.*, vol. 12, no. 4, pp. 173–187, 2000.

[65] A. Kumar, W. A. Edelstein, and P. A. Bottomley, "Noise figure limits for circular loop MR coils.," *Magn. Reson. Med.*, vol. 61, no. 5, pp. 1201–1209, May 2009.

[66] E. Fukushima, "Radiofrequency Coils for NMR: A Peripatetic History of Their Twists and Turns," in *RF Coils for MRI*, West Sussex: A John Wiley and Sons, Ltd., Publication, 2012, pp. 9–17.

[67] J. J. Ackerman, T. H. Grove, G. G. Wong, D. G. Gadian, and G. K. Radda, "Mapping of metabolites in whole animals by 31P NMR using surface coils.," *Nature*, vol. 283, no. 5743, pp. 167–170, Jan. 1980.

[68] J. Mispelter, M. Lupu, and A. Briguet, *NMR Probeheads for Biophysical and Biomedical Experiments: Theoretical Principles & Practical Guidelines*. Imperial College Press, 2006.

[69] W. Froncisz and J. S. Hyde, "The loop-gap resonator: a new microwave lumped circuit ESR sample structure," *J. Magn. Reson.*, vol. 47, no. 3, pp. 515–521, May 1982.

[70] É. É. N. Mascart, J. F. Joubert, and E. Atkinson, *A Treatise on Electricity and Magnetism*. T. De La Rue, 1883.

[71] C. Chin, C. M. Collins, S. Li, B. J. Dardzinski, and M. B. Smith, "BirdcageBuilder: Design of specified-geometry birdcage coils with desired current pattern and resonant frequency," *Concepts Magn. Reson.*, vol. 15, no. 2, pp. 156–163, Jun. 2002.

[72] G. H. Glover, C. E. Hayes, N. J. Pelc, W. A. Edelstein, O. M. Mueller, H. R. Hart, C. J. J. Hardy, M. O'Donnell, and W. D. D. Barber, "Comparison of linear and circular polarization for magnetic resonance imaging," *J. Magn. Reson.*, vol. 64, no. 2, pp. 255–270, Sep. 1985.

[73] J. Tropp, "The theory of the bird-cage resonator," *J. Magn. Reson.*, vol. 82, no. 1, pp. 51–62, Mar. 1989.

[74] E. K. Insko, M. A. Elliott, J. C. Schotland, and J. S. Leigh, "Generalized Reciprocity," *J. Magn. Reson.*, vol. 131, no. 1, pp. 111–117, Mar. 1998.

[75] R. Stollberger and P. Wach, "Imaging of the activeB1 fieldin vivo," *Magn. Reson. Med.*, vol. 35, no. 2, pp. 246–251, Feb. 1996.

[76] C. H. Cunningham, J. M. Pauly, and K. S. Nayak, "Saturated double-angle method for rapid B1+ mapping," *Magn. Reson. Med.*, vol. 55, no. 6, pp. 1326–1333, Jun. 2006.

[77] V. L. Yarnykh, "Actual flip-angle imaging in the pulsed steady state: A method for rapid three-dimensional mapping of the transmitted radiofrequency field," *Magn. Reson. Med.*, vol. 57, no. 1, pp. 192–200, Jan. 2007.

[78] G. R. Morrell, "A phase-sensitive method of flip angle mapping," *Magn. Reson. Med.*, vol. 60, no. 4, pp. 889–894, Oct. 2008.

[79] L. I. Sacolick, F. Wiesinger, I. Hancu, and M. W. Vogel, "B1 mapping by Bloch-Siegert shift," *Magn. Reson. Med.*, vol. 63, no. 5, pp. 1315–1322, May 2010.

[80] N. F. Ramsey, "Resonance Transitions Induced by Perturbations at Two or More Different Frequencies," *Phys. Rev.*, vol. 100, no. 4, pp. 1191–1194, Nov. 1955.

[81] E. M. Haacke, R. W. Brown, M. R. Thompson, and R. Venkatesan, *Magnetic resonance imaging: physical principles and sequence design*, 2nd ed. Wiley-Liss New York, 1999.

[82] D. P. Myer, "Radiofrequency Power Amplifiers for NMR and MRI," in *RF Coils for MRI*, West Sussex: John Wiley and Sons Ltd., 2012, pp. 299–314.

[83] D. I. Hoult, "Receiver Design for MR," in *RF Coils for MRI*, West Sussex: John Wiley and Sons Ltd., 2012, pp. 271–297.

[84] H. T. Friis, "Noise Figures of Radio Receivers," in *Proc. of the IRE*, 1944, pp. 419–422.

[85] D. Moskau, "Application of real time digital filters in NMR spectroscopy," *Concepts Magn. Reson.*, vol. 15, no. 2, pp. 164–176, 2002.

[86] M. A. Delsuc and J. Y. Lallemand, "Improvement of dynamic range in NMR by oversampling," *J. Magn. Reson.*, vol. 69, pp. 504–507, 1986.

[87] L. Devlin, "The design of integrated switches and phase shifters," in *COLLOQUIUM DIGEST-IEE*, 1999, p. 2/1-2/14.

[88] F. A. Duck, *Physical Properties of Tissue: A Comprehensive Reference Book*. San Diego, United States: Elsevier Science Publishing Co Inc, 1990.

[89] M. Paley, "MRI of tissues with short T2s and T2*s. G. M. Bydder, G. D. Fullerton and I. R. Young.," *NMR Biomed.*, vol. 26, no. 10, pp. 1336–1337, Oct. 2013.

[90] F. J. Rühli, R. K. Chhem, and T. Böni, "Diagnostic paleoradiology of mummified tissue: interpretation and pitfalls.," *Can. Assoc. Radiol. J.*, vol. 55, no. 4, pp. 218–27, Oct. 2004.

[91] F. J. Rühli, T. Böni, J. Perlo, F. Casanova, M. Baias, E. Egarter, and B. Blümich, "Non-invasive spatial tissue discrimination in ancient mummies and bones in situ by portable nuclear magnetic resonance," *J. Cult. Herit.*, vol. 8, no. 3, pp. 257–263, Jul. 2007.

[92] Y. Wu, J. L. Ackerman, D. A. Chesler, L. Graham, Y. Wang, and M. J. Glimcher, "Density of organic matrix of native mineralized bone measured by water- and fat-suppressed proton projection {MRI}," *Magn. Reson. Med.*, vol. 50, no. 1, pp. 59–68, 2003.

[93] U. Eliav and G. Navon, "A Study of Dipolar Interactions and Dynamic Processes of Water Molecules in Tendon by 1H and 2H Homonuclear and Heteronuclear Multiple-Quantum-Filtered NMR Spectroscopy," *J. Magn. Reson.*, vol. 137, no. 2, pp. 295–310, Apr. 1999.

[94] J. S. Nyman, Q. Ni, D. P. Nicolella, and X. Wang, "Measurements of mobile and bound water by nuclear magnetic resonance correlate with mechanical properties of bone," *Bone*, vol. 42, pp. 193–199, 2008.

[95] R. A. Horch, J. S. Nyman, D. F. Gochberg, R. D. Dortch, and M. D. Does, "Characterization of 1H NMR signal in human cortical bone for magnetic resonance imaging.," *Magn. Reson. Med.*, vol. 64, pp. 680–687, 2010.

[96] H. S. Rad, S. C. B. Lam, J. F. Magland, H. Ong, C. Li, H. K. Song, J. Love, and F. W. Wehrli, "Quantifying cortical bone water in vivo by three-dimensional ultra-short echo-time MRI," *NMR Biomed.*, vol. 24, no. 7, pp. 855–864, 2011.

[97] A. Techawiboonwong, H. K. Song, M. B. Leonard, and F. W. Wehrli, "Cortical Bone Water: In Vivo Quantification with Ultrashort Echo-Time MR Imaging," *Radiology*, vol. 248, no. 3, pp. 824–833, Sep. 2008.

[98] D. Capitani, V. Di Tullio, and N. Proietti, "Nuclear magnetic resonance to characterize and monitor Cultural Heritage.," *Prog. Nucl. Magn. Reson. Spectrosc.*, vol. 64, pp. 29–69, Jul. 2012.

[99] A. Lucas, "The Use of Natron in Mummification," *J. Egypt. Archaeol.*, vol. 18, no. 3/4, p. 125, Nov. 1932.

[100] A. T. Sandison, "The Use of Natron in Mummification in Ancient Egypt," *J. Near East. Stud.*, vol. 22, no. 4, p. 259, 1963.

[101] F. Herodot, *Historien 1/2. Zweisprachige Ausgabe: Griechisch-Deutsch.* Dusseldorf: Artemis & Winkler, 2000.

[102] A. C. Aufderheide, *The Scientific Study of Mummies*. Cambridge University Press, 2003.

[103] B. Adams, *Egyptian Mummies*. Shire Publications, 1988.

[104] S. A. Buckley, A. W. Stott, and R. P. Evershed, "Studies of organic residues from ancient Egyptian mummies using high temperature-gas chromatography-mass spectrometry and sequential thermal desorption-gas chromatography-mass spectrometry and pyrolysis-gas chromatography-mass spectrometry.," *Analyst*, vol. 124, pp. 443–452, 1999.

[105] S. A. Buckley and R. P. Evershed, "Organic chemistry of embalming agents in Pharaonic and Graeco-Roman mummies.," *Nature*, vol. 413, no. 6858, pp. 837–41, Oct. 2001.

[106] M. Perla Colombini, F. Modugno, F. Silvano, and M. Onor, "Characterization of the balm of an Egyptian mummy from the seventh century B.C.," *Stud. Conserv.*, vol. 45, no. 1, pp. 19–29, 2000.

[107] S. A. Buckley, K. A. Clark, and R. P. Evershed, "Complex organic chemical balms of Pharaonic animal mummies," *Nature*, vol. 431, no. 7006, pp. 294–299, Sep. 2004.

[108] M. J. Collins, C. M. Nielsen-Marsh, J. Hiller, C. I. Smith, J. P. Roberts, R. V. Prigodich, T. J. Wess, J. Csapo, a. R. Millard, and G. Turner-Walker, "The survival of organic matter in bone: a review," *Archaeometry*, vol. 44, no. 3, pp. 383–394, Aug. 2002.

[109] D. Alfano, A. R. Albunia, O. Motta, and A. Proto, "Detection of diagenetic alterations by Spectroscopic Analysis on Archaeological Bones from the Necropolis of Poseidonia (Paestum): A case study," *J. Cult. Herit.*, vol. 10, no. 4, pp. 509–513, Oct. 2009.

[110] R. E. M. Hedges, "Bone diagenesis: an overview of processes," *Archaeometry*, vol. 44, no. 3, pp. 319–328, Aug. 2002.

[111] H. Piepenbrink, J. Frahm, A. Haase, and D. Matthaei, "Nuclear magnetic resonance imaging of mummified corpses," *Am. J. Phys. Anthropol.*, vol. 70, no. 1, pp. 27–28, May 1986.

[112] D. H. Shin, I. S. Lee, M. J. Kim, C. S. Oh, J. B. Park, G. D. Bok, and D. S. Yoo, "Magnetic resonance imaging performed on a hydrated mummy of medieval Korea.," *J. Anat.*, vol. 216, no. 3, pp. 329–34, Mar. 2010.

[113] K. Münnemann, T. Böni, G. Colacicco, B. Blümich, and F. Rühli, "Noninvasive (1)H and (23)Na nuclear magnetic resonance imaging of ancient Egyptian human mummified tissue.," *Magn. Reson. Imaging*, vol. 25, no. 9, pp. 1341–5, Nov. 2007.

[114] F. J. Rühli, H. von Waldburg, S. Nielles-Vallespin, T. Böni, and P. Speier, "Clinical Magnetic Resonance Imaging of Ancient Dry Human Mummies Without Rehydration," *JAMA*, vol. 298, no. 22, p. 2617, Dec. 2007.

[115] L. M. Ohrstrom, H. von Waldburg, P. Speier, M. Bock, R. E. Suri, and F. J. Ruhli, "Scenes from the past: MR imaging versus CT of ancient Peruvian and Egyptian mummified tissues," *Radiographics*, vol. 33, no. 1, pp. 291–296, 2013.

[116] M. Bock, P. Speier, S. Nielles-vallespin, M. Szimtennings, K. Leotta, and F. J. Ruhli, "MRI of an Egyptian Mummy on Clinical 1.5 and 3 T Whole Body Imagers," in *Proc. Intl. Soc. Mag. Reson. Med.*, 2008, p. 3666.

[117] D. J. Tyler, M. D. Robson, R. M. Henkelman, I. R. Young, and G. M. Bydder, "Magnetic resonance imaging with ultrashort TE (UTE) PULSE sequences: Technical considerations," *J. Magn. Reson. Imaging*, vol. 25, no. 2, pp. 279–289, Feb. 2007.

[118] R. Biswas, W. Bae, E. Diaz, K. Masuda, C. B. Chung, G. M. Bydder, and J. Du, "Ultrashort echo time (UTE) imaging with bi-component analysis: Bound and free water evaluation of bovine cortical bone subject to sequential drying," *Bone*, vol. 50, no. 3, pp. 749–755, 2012.

[119] S. Emid and J. H. N. Creyghton, "High Resolution NMR Imaging in Solids," *Physica*, vol. 128B, pp. 81–83, 1985.

[120] D. M. Grodzki, P. M. Jakob, and B. Heismann, "Ultrashort echo time imaging using pointwise encoding time reduction with radial acquisition (PETRA).," *Magn. Reson. Med.*, vol. 67, no. 2, pp. 510–8, Feb. 2012.

[121] F. J. Rühli, H. von Waldburg, S. Nielles-Vallespin, T. Böni, and P. Speier, "Clinical Magnetic Resonance Imaging of Ancient Dry Human Mummies Without Rehydration," *JAMA*, vol. 298, no. 22, p. 2617, Dec. 2007.

[122] A. G. Siu, A. Ramadeen, X. Hu, L. Morikawa, L. Zhang, J. Y. C. Lau, G. Liu, M. Pop, K. A. Connelly, P. Dorian, and G. A. Wright, "Characterization of the ultrashort-TE (UTE) MR collagen signal," *NMR Biomed.*, vol. 28, no. 10, pp. 1236–1244, Oct. 2015.

[123] E. Y. Chang, J. Du, K. Iwasaki, R. Biswas, S. Statum, Q. He, W. C. Bae, and C. B. Chung, "Single- and Bi-component T2* analysis of tendon before and during tensile loading, using UTE sequences," *J. Magn. Reson. Imaging*, vol. 42, no. 1, pp. 114–120, Jul. 2015.

[124] J. Du, M. Carl, M. Bydder, A. Takahashi, C. B. Chung, and G. M. Bydder, "Qualitative and quantitative ultrashort echo time (UTE) imaging of cortical bone," *J. Magn. Reson.*, vol. 207, no. 2, pp. 304–311, Dec. 2010.

[125] M. D. Robson and G. M. Bydder, "Clinical ultrashort echo time imaging of bone and other connective tissues," *NMR Biomed.*, vol. 19, no. 7, pp. 765–780, 2006.

[126] S. Josan, J. M. Pauly, B. L. Daniel, and K. B. Pauly, "Double Half RF pulses for reduced sensitivity to eddy currents in UTE imaging," *Magn. Reson. Med.*, vol. 61, no. 5, pp. 1083–1089, 2009.

[127] K. D. Harkins, R. A. Horch, and M. D. Does, "Simple and robust saturation-based slice selection for ultrashort echo time MRI," *Magn. Reson. Med.*, vol. 73, no. 6, pp. 2204–2211, Jun. 2015.

[128] Y. Wu, G. Dai, J. L. Ackerman, M. I. Hrovat, M. J. Glimcher, B. D. Snyder, A. Nazarian, and D. a Chesler, "Water- and fat-suppressed proton projection MRI (WASPI) of rat femur bone," *Magn. Reson. Med.*, vol. 57, no. 3, pp. 554–567, Mar. 2007.

[129] B. J. Balcom, R. P. Macgregor, S. D. Beyea, D. P. Green, R. L. Armstrong, and T. W. Bremner, "Single-Point Ramped Imaging with T1 Enhancement (SPRITE)," *J. Magn. Reson. Ser. A*, vol. 123, no. 1, pp. 131–134, Nov. 1996.

[130] J. M. Pauly, S. M. Conolly, and D. G. Nishimura, "Magnetic resonance imaging of short T2 species with improved contrast," US5150053 A, 1992.

[131] A. M. El-Sharkawy, M. Schär, P. a Bottomley, and E. Atalar, "Monitoring and correcting spatio-temporal variations of the MR scanner's static magnetic field," *Magn. Reson. Mater. Physics, Biol. Med.*, vol. 19, no. 5, pp. 223–236, Nov. 2006.

[132] K. C. Chu and B. K. Rutt, "MR gradient coil heat dissipation," *Magn. Reson. Med.*, vol. 34, no. 1, pp. 125–132, Jul. 1995.

[133] Y. Chen and B. Rutt, "An Analytical Model of Gradient Coil Heating," in *Proc. Intl. Soc. Mag. Reson. Med. 11*, 2004, p. 1629.

[134] D. M. Grodzki, P. M. Jakob, and B. Heismann, "Correcting slice selectivity in hard pulse sequences," *J. Magn. Reson.*, vol. 214, pp. 61–67, 2012.

[135] M. Weiger, K. P. Pruessmann, and F. Hennel, "{MRI} with zero echo time: hard versus sweep pulse excitation," *Magn. Reson. Med.*, vol. 66, no. 2, pp. 379–389, 2011.

[136] R. M. Henkelman, "Measurement of signal intensities in the presence of noise in MR images," *Med. Phys.*, vol. 12, no. 2, p. 232, 1985.

[137] W. A. Edelstein, G. H. Glover, C. J. Hardy, and R. W. Redington, "The intrinsic signal-to-noise ratio in NMR imaging.," *Magn. Reson. Med.*, vol. 3, no. 4, pp. 604–618, Aug. 1986.

[138] D. A. Ortendahl, N. M. Hylton, L. Kaufman, and L. E. Crooks, "Signal to noise in derived NMR images," *Magn. Reson. Med.*, vol. 1, pp. 316–338, 1984.

[139] R. K. Gupta, "A new look at the method of variable nutation angle for the measurement of spin-lattice relaxation times using fourier transform NMR," *J. Magn. Reson.*, vol. 25, no. 1, pp. 231–235, Jan. 1977.

[140] F. Springer, G. Steidle, P. Martirosian, C. D. Claussen, and F. Schick, "Effects of in-pulse transverse relaxation in 3D ultrashort echo time sequences: Analytical derivation, comparison to numerical simulation and experimental application at 3T," *J. Magn. Reson.*, vol. 206, no. 1, pp. 88–96, Sep. 2010.

[141] Y. Eryaman, B. Akin, and E. Atalar, "Reduction of implant RF heating through modification of transmit coil electric field," *Magn. Reson. Med.*, vol. 65, no. 5, pp. 1305–1313, 2011.

[142] A. C. Özen, M. Bock, and E. Atalar, "Active decoupling of RF coils using a transmit array system," *Magn. Reson. Mater. Physics, Biol. Med.*, vol. 28, no. 6, pp. 565–576, Dec. 2015.

[143] A. C. Özen, E. Atalar, J. Korvink, and M. Bock, "In vivo Concurrent Excitation and Acquisition MRI with Self-referenced Active Decoupling," in *Proc. Intl. Soc. Mag. Reson. Med. 24*, 2016, p. 2179.

[144] A. Sahai, G. Patel, C. Dick, and A. Sabharwal, "On the impact of phase noise on active cancelation in wireless full-duplex," *IEEE Trans. Veh. Technol.*, vol. 62, no. 9, pp. 4494–4510, 2013.

[145] J. Il Choi, M. Jain, K. Srinivasan, P. Levis, and S. Katti, "Achieving single channel, full duplex wireless communication," in *Proceedings of the sixteenth annual international conference on Mobile computing and networking - MobiCom '10*, 2010, pp. 1–12.

[146] E. Everett, A. Sahai, and A. Sabharwal, "Passive self-interference suppression for full-duplex infrastructure nodes," *IEEE Trans. Wirel. Commun.*, vol. 13, no. 2, pp. 680–694, 2014.

[147] M. E. Knox, "Single antenna full duplex communications using a common carrier," in *WAMICON 2012 IEEE Wireless & Microwave Technology Conference*, 2012, pp. 1–6.

[148] M. Duarte, C. Dick, and A. Sabharwal, "Experiment-driven characterization of full-duplex wireless systems," *IEEE Trans. Wirel. Commun.*, vol. 11, no. 12, pp. 4296–4307, 2012.

[149] D. Bharadia, E. McMilin, and S. Katti, "Full duplex radios," in *Proceedings of the ACM SIGCOMM 2013 conference on SIGCOMM - SIGCOMM '13*, 2013, vol. 43, no. 4, p. 375.

[150] M. Salim, A. C. Özen, M. Bock, and E. Atalar, "Detection of MR Signal during RF Excitation using Full-Duplex Radio System," in *Proc. Intl. Soc. Mag. Reson. Med. 24*, 2016, p. 3636.

[151] S.-M. Sohn, J. T. Vaughan, M. Garwood, and D. Idiyatulli, "The first demonstration of simultaneous transmit and receive MRI in vivo," in *Proc. Intl. Soc. Mag. Reson. Med. 24*, 2016, p. 542.

[152] Microsemi, "The PIN Diode Circuit Designer's Handbook," Watertown, 1998.

[153] D. M. Pozar, *Microwave engineering*. John Wiley Sons, 2009.

[154] D. Idiyatullin, C. Corum, S. Moeller, and M. Garwood, "Gapped pulses for frequency-swept MRI," *J. Magn. Reson.*, vol. 193, no. 2, pp. 267–273, Aug. 2008.

[155] R. R. Ernst, "Magnetic resonance with stochastic excitation," *J. Magn. Reson.*, vol. 3, no. 1, pp. 10–27, Jul. 1970.

[156] M. Hirano, "Vocal mechanisms in singing: Laryngological and phoniatric aspects," *J. Voice*, vol. 2, no. 1, pp. 51–69, Aug. 1988.

[157] I. R. Titze, *Principles of Voice Production*. Prentice Hall, 1994.

[158] P. Kitzing, "Clinical applications of electroglottography," *J. Voice*, vol. 4, no. 3, pp. 238–249, Aug. 1990.

[159] N. Henrich, C. d'Alessandro, B. Doval, and M. Castellengo, "On the use of the derivative of electroglottographic signals for characterization of nonpathological phonation.," *J. Acoust. Soc. Am.*, vol. 115, no. 3, pp. 1321–1332, Mar. 2004.

[160] M. Echternach, L. Traser, and B. Richter, "Perturbation of Voice Signals in Register Transitions on Sustained Frequency in Professional Tenors," *J. Voice*, vol. 26, no. 5, p. 674.e9-674.e15, Sep. 2012.

[161] M. Echternach and B. Richter, "Passaggio in the professional tenor voice--evaluation of perturbation measures.," *J. Voice*, vol. 26, no. 4, pp. 440–446, Jul. 2012.

[162] M. Rothenberg and J. J. Mahshie, "Monitoring vocal fold abduction through vocal fold contact area.," *J. Speech Hear. Res.*, vol. 31, no. 3, pp. 338–351, Sep. 1988.

[163] R. H. Colton, J. K. Casper, and R. Leonard, *Understanding Voice Problems: A Physiological Perspective for Diagnosis and Treatment*. Lippincott Williams & Wilkins, 2006.

[164] A. Ziethe, R. Patel, M. Kunduk, U. Eysholdt, and S. Graf, "Clinical Analysis Methods of Voice Disorders," *Curr. Bioinform.*, vol. 6, no. 3, pp. 270–285, Sep. 2011.

[165] J. Sundberg, *The Science of the Singing Voice*. Northern Illinois University Press, 1987.

[166] M. Echternach, M. Markl, and B. Richter, "Dynamic real-time magnetic resonance imaging for the analysis of voice physiology.," *Curr. Opin. Otolaryngol. Head Neck Surg.*, vol. 20, no. 6, pp. 450–457, Dec. 2012.

[167] J. M. Wild, H. Marshall, M. Bock, L. R. Schad, P. M. Jakob, M. Puderbach, F. Molinari, E. J. R. van Beek, and J. Biederer, "MRI of the lung (1/3): Methods," *Insights Imaging*, vol. 3, no. 4, pp. 345–353, 2012.

[168] J. Biederer, M. Beer, W. Hirsch, J. Wild, M. Fabel, M. Puderbach, and E. J. R. van Beek, "MRI of the lung (2/3). Why... when ... how?," *Insights Imaging*, vol. 3, no. 4, pp. 355–371, 2012.

[169] J. Biederer, S. Mirsadraee, M. Beer, F. Molinari, C. Hintze, G. Bauman, M. Both, E. J. R. Van Beek, J. Wild, and M. Puderbach, "MRI of the lung (3/3)—current applications and future perspectives," *Insights Imaging*, vol. 3, no. 4, pp. 373–386, Aug. 2012.

[170] C. Plathow, M. Schoebinger, C. Fink, S. Ley, M. Puderbach, M. Eichinger, M. Bock, H.-P. Meinzer, and H.-U. Kauczor, "Evaluation of lung volumetry using dynamic three-dimensional magnetic resonance imaging.," *Invest. Radiol.*, vol. 40, no. 3, pp. 173–179, Mar. 2005.

[171] M. Echternach, L. Traser, M. Markl, and B. Richter, "Vocal tract configurations in male alto register functions.," *J. Voice*, vol. 25, no. 6, pp. 670–677, Nov. 2011.

[172] A. D. Scott, R. Boubertakh, M. J. Birch, and M. E. Miquel, "Towards clinical assessment of velopharyngeal closure using MRI: evaluation of real-time MRI sequences at 1.5 and 3 T," *Br. J. Radiol.*, vol. 85, no. 1019, pp. e1083–e1092, Nov. 2012.

[173] M. Echternach, L. Traser, and B. Richter, "Vocal tract configurations in tenors' passaggio in different vowel conditions-a real-time magnetic resonance imaging study.," *J. Voice*, vol. 28, no. 2, p. 262.e1-262.e8, Mar. 2014.

[174] S. Narayanan, K. Nayak, S. Lee, A. Sethy, and D. Byrd, "An approach to real-time magnetic resonance imaging for speech production.," *J. Acoust. Soc. Am.*, vol. 115, no. 4, pp. 1771–1776, Apr. 2004.

[175] Y.-C. Kim, C. E. Hayes, S. S. Narayanan, and K. S. Nayak, "Novel 16-channel receive coil array for accelerated upper airway MRI at 3 Tesla.," *Magn. Reson. Med.*, vol. 65, no. 6, pp. 1711–1717, Jun. 2011.

[176] B. P. Sutton, C. A. Conway, Y. Bae, R. Seethamraju, and D. P. Kuehn, "Faster dynamic imaging of speech with field inhomogeneity corrected spiral fast low angle shot (FLASH) at 3 T.," *J. Magn. Reson. Imaging*, vol. 32, no. 5, pp. 1228–1237, Nov. 2010.

[177] Y. Bae, D. P. Kuehn, C. A. Conway, and B. P. Sutton, "Real-time magnetic resonance imaging of velopharyngeal activities with simultaneous speech recordings.," *Cleft Palate. Craniofac. J.*, vol. 48, no. 6, pp. 695–707, Nov. 2011.

[178] A. Niebergall, S. Zhang, E. Kunay, G. Keydana, M. Job, M. Uecker, and J. Frahm, "Real-time MRI of speaking at a resolution of 33 ms: Undersampled radial FLASH with nonlinear inverse reconstruction," *Magn. Reson. Med.*, vol. 69, no. 2, pp. 477–485, 2013.

[179] M. Burdumy, L. Traser, B. Richter, M. Echternach, J. G. Korvink, J. Hennig, and M. Zaitsev, "Acceleration of MRI of the vocal tract provides additional insight into articulator modifications," *J. Magn. Reson. Imaging*, vol. 42, no. 4, pp. 925–935, Oct. 2015.

[180] A. C. Özen, U. Ludwig, L. M. Öhrström, F. J. Rühli, and M. Bock, "Comparison of ultrashort echo time sequences for MRI of an ancient mummified human hand," *Magn. Reson. Med.*, vol. 75, no. 2, pp. 701–708, Feb. 2016.

[181] A. C. Özen, L. Traser, M. Echternach, T. Dadakova, M. Burdumy, B. Richter, and M. Bock, "Ensuring safety and functionality of electroglottography measurements during dynamic pulmonary MRI," *Magn. Reson. Med.*, vol. 76, no. 5, pp. 1629–1635, Nov. 2016.

[182] K. Block and M. Uecker, "Simple method for adaptive gradient-delay compensation in radial MRI," *Proc. Int. Soc. Magn. Reson. Imaging*, vol. 19, p. 2816, 2011.

[183] A. C. Özen, U. Ludwig, and M. Bock, "' Be nice to your gradients !' Useful gradient modifications for SPI," in *Proceedings of Joint Annual Meeting ISMRM-ESMRMB*, 2014, p. 4251.

[184] Y. Jung, Y. Jashnani, R. Kijowski, and W. F. Block, "Consistent non-cartesian off-axis MRI quality: Calibrating and removing multiple sources of demodulation phase errors," *Magn. Reson. Med.*, vol. 57, no. 1, pp. 206–212, Jan. 2007.

[185] J. G. Raya, O. Dietrich, A. Horng, J. Weber, M. F. Reiser, and C. Glaser, "T2 measurement in articular cartilage: impact of the fitting method on accuracy and precision at low SNR," *Magn. Reson. Med.*, vol. 63, no. 1, pp. 181–193, Jan. 2010.

[186] A. J. Miller and P. M. Joseph, "The use of power images to perform quantitative analysis on low SNR MR images.," *Magn. Reson. Imaging*, vol. 11, no. 7, pp. 1051–1056, 1993.

[187] R. F. Lee, R. O. Giaquinto, and C. J. Hardy, "Coupling and decoupling theory and its application to the MRI phased array," *Magn. Reson. Med.*, vol. 48, no. 1, pp. 203–213, Jul. 2002.

[188] R. Ludwig and G. Bogdanov, *RF Circuit Design: Theory and Applications*. Prentice-Hall, 2009.

[189] U. Klose, "Mapping of the radio frequency magnetic field with a MR snapshot FLASH technique.," *Med. Phys.*, vol. 19, no. 4, pp. 1099–1104, 1992.

[190] J. I. Jackson, C. H. Meyer, D. G. Nishimura, and A. Macovski, "Selection of a convolution function for {F}ourier inversion using gridding," *Med. Imaging, IEEE Trans.*, vol. 10, no. 3, pp. 473–478, 1991.

[191] J. P. Starski, "Correction to 'Optimization of the Matching Network for a Hybrid Coupler Phase Shifter' (Letters)," *IEEE Trans. Microw. Theory Tech.*, vol. 26, no. 4, pp. 662–666, Apr. 1978.

[192] S. Inati, "IceNIH_RawSend." Siemens, IDEA, 2014.

[193] A. C. Özen, J. Korvink, and M. Bock, "Concurrent Excitation and Acquisition in Steady State: T1-Modulation Effects of Frequency Sweep," in *Proc. Intl. Soc. Mag. Reson. Med. 24*, 2016, p. 1809.

[194] D. G. Nishimura, *Principles of magnetic resonance imaging*. Stanford University, 1996.

[195] F. G. Shellock, "Monitoring during MRI. An evaluation of the effect of high-field MRI on various patient monitors.," *Med. Electron.*, vol. 17, no. 4, pp. 93–97, Sep. 1986.

[196] B. A. Holshouser, D. B. J. Hinshaw, and F. G. Shellock, "Sedation, anesthesia, and physiologic monitoring during MR imaging: evaluation of procedures and equipment.," *J. Magn. Reson. Imaging*, vol. 3, no. 3, pp. 553–558, 1993.

[197] A. von Smekal, K. C. Seelos, C. R. Küper, and M. Reiser, "Patient monitoring and safety during MRI examinations," *Eur. Radiol.*, vol. 5, no. 3, pp. 302–305, 1995.

[198] P. Lanzer, C. Barta, E. H. Botvinick, H. U. Wiesendanger, G. Modin, and C. B. Higgins, "ECG-synchronized cardiac MR imaging: method and evaluation.," *Radiology*, vol. 155, no. 3, pp. 681–686, Jun. 1985.

[199] J. Felblinger, C. Lehmann, and C. Boesch, "Electrocardiogram acquisition during MR examinations for patient monitoring and sequence triggering.," *Magn. Reson. Med.*, vol. 32, no. 4, pp. 523–529, Oct. 1994.

[200] R. E. Wendt, R. Rokey, G. W. Vick, and D. L. Johnston, "Electrocardiographic gating and monitoring in NMR imaging.," *Magn. Reson. Imaging*, vol. 6, no. 1, pp. 89–95, 1988.

[201] S. Warach, J. R. Ives, G. Schlaug, M. R. Patel, D. G. Darby, V. Thangaraj, R. R. Edelman, and D. L. Schomer, "EEG-triggered echo-planar functional MRI in epilepsy.," *Neurology*, vol. 47, no. 1, pp. 89–93, Jul. 1996.

[202] J. R. Ives, S. Warach, F. Schmitt, R. R. Edelman, and D. L. Schomer, "Monitoring the patient's EEG during echo planar MRI.," *Electroencephalogr. Clin. Neurophysiol.*, vol. 87, no. 6, pp. 417–420, Dec. 1993.

[203] J. Jacobs, J. Stich, B. Zahneisen, J. Assländer, G. Ramantani, A. Schulze-Bonhage, R. Korinthenberg, J. Hennig, and P. LeVan, "Fast fMRI provides high statistical power in the analysis of epileptic networks," *Neuroimage*, vol. 88, pp. 282–294, 2014.

[204] J. Felblinger, R. M. Müri, C. Ozdoba, G. Schroth, C. W. Hess, and C. Boesch, "Recordings of eye movements for stimulus control during fMRI by means of electro-oculographic methods.," *Magn. Reson. Med.*, vol. 36, no. 3, pp. 410–414, 1996.

[205] D. A. Seeber, J. Jevtic, and A. Menon, "Floating shield current suppression trap," *Concepts Magn. Reson. Part B Magn. Reson. Eng.*, vol. 21B, no. 1, pp. 26–31, 2004.

[206] P. Comon, "Independent component analysis, A new concept?," *Signal Processing*, vol. 36, no. 3, pp. 287–314, 1994.

[207] A. Hyvärinen, "A fast fixed-point algorithm for Independent Component Analysis," *Neural Comput.*, vol. 9, pp. 1483–1492, 1997.

[208] A. Hyvarinen, "New approximations of differential entropy for independent component analysis and projection pursuit," *Adv. Neural Inf. Process. Syst.*, vol. 10, pp. 273–279, 1998.

[209] A. Hyvarinen, "Fast and robust fixed-point algorithms for independent component analysis," *IEEE Trans. Neural Networks*, vol. 10, no. 3, pp. 626–634, May 1999.

[210] A. Hyvärinen and E. Oja, "Independent component analysis: algorithms and applications," *Neural Networks*, vol. 13, no. 4–5, pp. 411–430, Jun. 2000.

[211] C. J. James and C. W. Hesse, "On the use of Spectrally Constrained ICA applied to single-channel Ictal EEG recordings within a Dynamical Embedding Framework," in *Proc. IEEE Engineering in Medicine and Biology 27th Annual Conference*, 2005, vol. 1, pp. 956–959.

[212] American Society for Testing and Materials International, "Designation: ASTM F2052-06e1, standard test method for measurement of magnetically induced displacement force on medical devices in the magnetic resonance environment," West Conshohocken, PA, 2006.

[213] American Society for Testing and Materials International, "Designation: ASTM F2213-06, standard test method for measurement of magnetically induced torque on medical devices in the magnetic resonance environment," West Conshohocken, PA, 2011.

[214] American Society for Testing and Materials International, "Designation: ASTM F2119-07, standard test method for evaluation of MR image artifacts from passive implants.," West Conshohocken, PA, 2013.

[215] American Society for Testing and Materials International, "Designation: ASTM F2182-11a, standard test method for measurement of radio frequency induced heating near passive implants during magnetic resonance imaging," West Conshohocken, PA, 2011.

[216] A. Galloway, W. H. Birkby, A. M. Jones, T. E. Henry, and B. O. Parks, "Decay rates of human remains in an arid environment," *J Forensic Sci*, vol. 34, no. 3, pp. 607–616, 1989.

[217] F. Schulz, M. Tsokos, and K. Püschel, "Natürliche Mumifikation im häuslichen Milieu," *Rechtsmedizin*, vol. 10, no. 1, pp. 32–38, Dec. 1999.

[218] R. W. Byard and M. Tsokos, "Forensic Issues in Cases of Diogenes Syndrome," *Am. J. Forensic Med. Pathol.*, vol. 28, no. 2, pp. 177–181, Jun. 2007.

[219] C. P. Campobasso, R. Falamingo, I. Grattagliano, and F. Vinci, "The Mummified Corpse in a Domestic Setting," *Am. J. Forensic Med. Pathol.*, vol. 30, no. 3, pp. 307–310, Sep. 2009.

[220] M. Weiger, D. O. Brunner, B. E. Dietrich, C. F. Müller, and K. P. Pruessmann, "ZTE imaging in humans," *Magn. Reson. Med.*, vol. 70, no. 2, pp. 328–332, Aug. 2013.

[221] M. Halse, J. Rioux, S. Romanzetti, J. Kaffanke, B. MacMillan, I. Mastikhin, N. J. Shah, E. Aubanel, and B. J. Balcom, "Centric scan SPRITE magnetic resonance imaging: optimization of SNR, resolution, and relaxation time mapping," *J. Magn. Reson.*, vol. 169, no. 1, pp. 102–117, Jul. 2004.

[222] R. A. De Graaf, *In vivo NMR spectroscopy: principles and techniques*. Wiley-Interscience, 2008.

[223] A. Hajimiri and T. H. Lee, "A general theory of phase noise in electrical oscillators," *IEEE J. Solid-State Circuits*, vol. 33, no. 2, pp. 179–194, 1998.

[224] T. H. Lee and A. Hajimiri, "Oscillator Phase Noise: A Tutorial," *IEEE J. Solid-State Circuits*, vol. 35, no. 3, pp. 326–336, 2000.

[225] I. Bahl and P. Bhartia, *Microwave solid state circuit design*. New Jersey: Wiley-Interscience, 2003.

[226] B. Kaa, "A simple approach to YIG oscillators," *VHF Commun*, vol. 4, pp. 217–224, 2004.

[227] T. Mimura, "The early history of the high electron mobility transistor (HEMT)," *IEEE Trans. Microw. Theory Tech.*, vol. 50, no. 3, pp. 780–782, 2002.

[228] I. Graesslin, D. Glaesel, P. Börnert, H. Stahl, P. Koken, K. Nehrke, H. Dingemans, G. Mens, J. Götze, and P. Harvey, "An Alternative Concept of Non-sequence-interfering Patient Respiration Monitoring," in *Proc. Intl. Soc. Mag. Reson. Med. 16*, 2008, p. 202.

[229] I. Graesslin, H. Stahl, K. Nehrke, P. Harvey, J. Smink, G. Mens, A. Senn, and P. Börnert, "An Alternative Concept for Non-Sequence Interfering, Contact-free Respiration Monitoring," in *Proc. Intl. Soc. Mag. Reson. Med. 17*, 2009, p. 752.

[230] P. P. Stang, S. M. Conolly, J. M. Santos, J. M. Pauly, and G. C. Scott, "Medusa: A Scalable MR Console Using USB," *IEEE Trans. Med. Imaging*, vol. 31, no. 2, pp. 370–379, Feb. 2012.

[231] I. Graesslin, S. Krueger, P. Vernickel, J. Achtzehn, K. Nehrke, and S. Weiss, "Detection of RF unsafe devices using a parallel transmission MR system," *Magn. Reson. Med.*, vol. 70, no. 5, pp. 1440–1449, Nov. 2013.

[232] J. Rahmer, P. Bornert, J. Groen, and C. Bos, "Three-dimensional radial ultrashort echo time imaging with {T_2} adapted sampling," *Magn. Reson. Med.*, vol. 55, no. 5, pp. 1075–1082, 2006.

[233] D. I. Hoult, G. Kolansky, D. Kripiakevich, and S. B. King, "The NMR multi-transmit phased array: a Cartesian feedback approach," *J. Magn. Reson.*, vol. 171, no. 1, pp. 64–70, Nov. 2004.

[234] A. C. Özen and E. Atalar, "Decoupling of Tx/Rx coils using a Tx-array System: Application to UTE and CEA," in *Proc. Intl. Soc. Mag. Reson. Med. 21*, 2013, p. 767.

[235] C. L. G. Ham, J. M. L. Engels, G. T. van de Wiel, and A. Machielsen, "Peripheral nerve stimulation during MRI: Effects of high gradient amplitudes and switching rates," *J. Magn. Reson. Imaging*, vol. 7, no. 5, pp. 933–937, Sep. 1997.

[236] R. E. Collin, *Foundations of Microwave Engineering*, 2nd ed. New York, USA: John Wiley and Sons, Inc., 2001.

[237] F. G. Shellock, *Reference manual for magnetic resonance safety, implants and devices: 2008 edition*. Los Angeles, CA: Biomedical Research Publishing Group, 2008.

[238] International Commission on Non-Ionising Radiation (ICNIRP), "Guidelines for limiting exposure for time varying electric, magnetic and electromagnetic fields," 1998.

[239] US Department of Health and Human Services Food and Drug Administration Center for Devices and Radiological Health, "Guidance for Industry and FDA Staff. Criteria for significant risk investigations of magnetic resonance diagnostic devices," Atlanta, GA, 2003.

[240] F. G. Shellock, T. O. Woods, and J. V Crues, "MR Labeling Information for Implants and Devices: Explanation of Terminology," *Radiology*, vol. 253, no. 1, pp. 26–30, Oct. 2009.

[241] F. G. Shellock and J. V Crues, "MR procedures: biologic effects, safety, and patient care," *Radiology*, vol. 232, pp. 635–652, 2004.

[242] J. E. Holmes and G. M. Bydder, "MR imaging with ultrashort TE (UTE) pulse sequences: Basic principles," *Radiography*, vol. 11, no. 3, pp. 163–174, Aug. 2005.

List of Figures

List of Tables

Acknowledgements

I would like to thank several people for making the completion of this thesis possible. First and foremost, I am grateful for the continuous support that my supervisor Prof. Dr. Michael Bock gave me during the entire process of my dissertation. He has not only guided me throughout the technical challenges, but also enlightened me on all aspects of academic research. I owe a special debt to him for his encouragement and all his constructive comments – although a bit harsh and direct at times. I really appreciated getting his critique since it enabled me to take my research one step further. He has always been generous with respect to his time as an advisor. I am proud and honored to be a part of his research team, which has been very productive and broadened my perspective.

I would like to express my gratitude to Prof. Dr. Jan Korvink for being interested in my research in the first place and stimulating insightful discussions. I am very grateful for having hosted me to conduct experiments in his lab together with Dr. Mazin Jouda and for having already started to realize some of the ideas that came out of our conversations. I am also grateful to him for examining my thesis.

I also owe special thanks to Prof. Dr. Michael Hoffmann for having accepted to be the chairman of my defense.

My deepest appreciation goes to Prof. Dr. Jürgen Hennig for having established a great research facility within the University Medical Center, Freiburg. I have benefited from all the facilities to a great extent conducting MRI experiments day and night.

I would also like to gratefully acknowledge Prof. Dr. Ergin Atalar for the engineering background I built during my masters study with him and for our ongoing collaborations.

A special thanks goes to Tracey Webb-Kolbinger for always helping me with all the administrative work.

Among the friends and colleagues who helped me along the way, I would like to single out the Experimental Radiology group members: Tetiana Dadakova, Dmitry Kurzhunov, Simon Reiss, Robert Borowiak, Dr. Axel Krafft, Dr. Ute Ludwig, and our old colleague Dr. John Matt Pavlina.

I wish to thank our collaborators from Department of Musicians' Medicine for the fruitful research experience, especially Dr. Louisa Traser.

I would also like to warmly thank PD. Dr. Jan Bernd Hoevener for having proofread a part of my thesis.

I owe a special debt to Dr. Emre Kopanoğlu for years of friendship and having proofread a part of my thesis.

All the years with intensive work, under the pressure of deadlines are only bearable with friends, family, and music.

We have been sharing the same office since I was an intern in National Magnetic Resonance Research Center in Ankara. I should acknowledge the ongoing conversations with him about an unlimited number of various topics: from science to politics, from plants to stars, from the meaning of life to optimization of the yoghurt fermentation, etc. Burak Akin is a very special friend and colleague. Together with Cem Uran, they have made Freiburg a more enjoyable place for me.

Thanks also goes to Dr. Elias Kellner. I can share one of the most valuable things for me with him: Music. Thanks to him I could get into the amazing world of Flamenco music with my dear Bağlama.

Speaking of music, and my instrument, I would like to acknowledge a very special friend Bülent Can, who inspires me a lot and makes me see the life from a different perspective.

I would like to express my gratitude to my cousin Talip Coşkun and his family for always being there when needed, for making me feel at home in Freiburg, and for giving me the opportunity of witnessing the growing up of my nephew and niece, Utku and Tuana.

Deepest thanks also goes to Christel and Joseph Gaida for having become a part of our family.

I also want to take a moment to thank deeply my high school Turkish literature teacher Ugur Erdem for being my life-long mentor.

I am eternally grateful to my parents Semra and Hidayet Özen and my brother Hasan for their unfailing and patient support throughout my life. The memories of our time keep my heart warm. They have selflessly encouraged me to explore new directions in life and seek my own destiny.

I would not believe that my PhD years would be the happiest years of my life. I am in debt to my loving wife Seval simply for making me a happier person.

List of Publications

Journal Papers

*Kurzhunov D, et al., Özen AC, Bock M. 3D CMRO2 mapping in human brain with direct 17O MRI: comparison of conventional and proton-constrained reconstructions. Neuroimage [Revision submitted, Feb 2017]

*Dadakova T, Krafft A, Özen AC, Bock M. Optimization of Acoustic Radiation Force Imaging: Influence of Timing Parameters on Sensitivity. Magn Reson Med. [in print]

*Traser L, Özen AC, et al. Respiratory dynamics in phonation and breathing- A real-time MRI study (2017) Respiratory Physiology & Neurobiology, 236:69-77.

*Özen AC, Traser L, Echternach M, Dadakova T, Burdumy M, Richter B and Bock M. (2016), Ensuring safety and functionality of electroglottography measurements during dynamic pulmonary MRI. Magn. Reson. Med. 76:1629–1635.

*Özen AC, Bock M, Atalar E. (2015) Active decoupling of RF coils using a transmit array system. Magnetic Resonance Materials in Physics, Biology and Medicine 28(6):565-576.

*Özen AC, Ludwig U, Öhrström LM, Rühli FJ and Bock M. (2016), Comparison of ultrashort echo time sequences for MRI of an ancient mummified human hand. Magn. Reson. Med. 75:701–708.

Patent Applications

***Özen AC**, Bock M. An Automated Active Analog Cancellation System with Dynamic Feedback for Concurrent Excitation and Acquisition in Magnetic Resonance. European Patent Application (Submitted – 2016 Aug 1)

Conference Papers

*__Özen AC__, Korvink J, Atalar E, Bock M (2016) In Vivo Concurrent Excitation and Acquisition MRI with Self-Referenced Active Decoupling. ISMRM 2016, Singapore. p 2179

*__Özen AC__, Korvink J, Bock M (2016) Concurrent Excitation and Acquisition in Steady State: T1-Modulation Effects of Frequency Sweep. ISMRM 2016, Singapore. p 1809

*Salim M, **Özen AC**, Bock M, Atalar E (2016) Detection of MR Signal During Rf Excitation Using Full Duplex Radio System. ISMRM 2016, Singapore. p 3636

*Dadakova T, **Özen AC**, Krafft A, Korvink J, Bock M (2016) Phase Artifact Correction for Improved ARFI Displacement Mapping Using a Short FID Navigator. ISMRM 2016, Singapore. p 3611

*__Özen AC__, et al. (2015) Active Decoupling of RF Coils: Application to 3D MRI with Concurrent Excitation and Acquisition. In: Proceedings of the 32nd ESMRMB Annual Meeting, Edinburgh, UK.

*__Özen AC__, et al. (2015) Which Technique is Best? Optimizing MRI of an Ancient Mummified Human Hand. In: Proceedings of the Evolutionary Medicine Conference, Zurich, Switzerland.

*Özen AC, et al. (2015) Ensuring Safety and Functionality of Electroglottography Measurements During Lung MRI. ISMRM 2015,Toronto.

*Özen AC, et al. (2015) Active Decoupling of RF Coils: Application to 3D MRI with Concurrent Excitation and Acquisition. ISMRM 2015,Toronto.

*Dadakova T, Özen AC, et al. (2015) MR-ARFI for the Quantification of Tissue Elastic Properties ISMRM 2015,Toronto.

*Özen AC, et al. (2014) Comparison of three ultrashort echo time sequences in MRI of an ancient Mummified hand. In: Proceedings of the DGMP Annual Meeting, Zurich, Switzerland

*Özen AC, et al. (2014) Comparison of UTE, PETRA and SPI sequences in MRI of Ancient Remains. ISMRM 2014, Milano.

*Özen AC, et al. (2014) "Be nice to your gradients!" Useful gradient modifications for SPI. ISMRM, Milano.

*Ludwig U, Özen AC, et al. (2014) ZTE MRI enables imaging of Egyptian Mummy: A comparison to CT and THz imaging. ISMRM, Milano.

*Pavlina JM, Özen AC, et al. (2014) Conductivity Effects on RF Surface Coils used for MR-Guided HIFU of the Prostate. ISMRM, Milano.

*Demir T, Özen AC, et al. (2014) Cylindrical Encoding in MRI. ISMRM, Milano

Schriften des Instituts für Mikrostrukturtechnik
am Karlsruher Institut für Technologie (KIT)

ISSN 1869-5183

Herausgeber: Institut für Mikrostrukturtechnik

Band 1 **Georg Obermaier**
Research-to-Business Beziehungen: Technologietransfer durch Kommunikation von Werten (Barrieren, Erfolgsfaktoren und Strategien). 2009
ISBN 978-3-86644-448-5

Band 2 **Thomas Grund**
Entwicklung von Kunststoff-Mikroventilen im Batch-Verfahren. 2010
ISBN 978-3-86644-496-6

Band 3 **Sven Schüle**
Modular adaptive mikrooptische Systeme in Kombination mit Mikroaktoren. 2010
ISBN 978-3-86644-529-1

Band 4 **Markus Simon**
Röntgenlinsen mit großer Apertur. 2010
ISBN 978-3-86644-530-7

Band 5 **K. Phillip Schierjott**
Miniaturisierte Kapillarelektrophorese zur kontinuierlichen Überwachung von Kationen und Anionen in Prozessströmen. 2010
ISBN 978-3-86644-523-9

Band 6 **Stephanie Kißling**
Chemische und elektrochemische Methoden zur Oberflächenbearbeitung von galvanogeformten Nickel-Mikrostrukturen. 2010
ISBN 978-3-86644-548-2

Die Bände sind unter www.ksp.kit.edu als PDF frei verfügbar oder als Druckausgabe zu bestellen.

Band 7 Friederike J. Gruhl
Oberflächenmodifikation von Surface Acoustic Wave (SAW) Biosensoren für biomedizinische Anwendungen. 2010
ISBN 978-3-86644-543-7

Band 8 Laura Zimmermann
Dreidimensional nanostrukturierte und superhydrophobe mikrofluidische Systeme zur Tröpfchengenerierung und -handhabung. 2011
ISBN 978-3-86644-634-2

Band 9 Martina Reinhardt
Funktionalisierte, polymere Mikrostrukturen für die dreidimensionale Zellkultur. 2011
ISBN 978-3-86644-616-8

Band 10 Mauno Schelb
Integrierte Sensoren mit photonischen Kristallen auf Polymerbasis. 2012
ISBN 978-3-86644-813-1

Band 11 Daniel Auernhammer
Integrierte Lagesensorik für ein adaptives mikrooptisches Ablenksystem. 2012
ISBN 978-3-86644-829-2

Band 12 Nils Z. Danckwardt
Pumpfreier Magnetpartikeltransport in einem Mikroreaktionssystem: Konzeption, Simulation und Machbarkeitsnachweis. 2012
ISBN 978-3-86644-846-9

Band 13 Alexander Kolew
Heißprägen von Verbundfolien für mikrofluidische Anwendungen. 2012
ISBN 978-3-86644-888-9

Band 14 Marko Brammer
Modulare Optoelektronische Mikrofluidische Backplane. 2012
ISBN 978-3-86644-920-6

Band 15 Christiane Neumann
Entwicklung einer Plattform zur individuellen Ansteuerung von Mikroventilen und Aktoren auf der Grundlage eines Phasenübergangs zum Einsatz in der Mikrofluidik. 2013
ISBN 978-3-86644-975-6

Band 16 Julian Hartbaum
Magnetisches Nanoaktorsystem. 2013
ISBN 978-3-86644-981-7

Band 17 Johannes Kenntner
Herstellung von Gitterstrukturen mit Aspektverhältnis 100 für die Phasenkontrastbildgebung in einem Talbot-Interferometer. 2013
ISBN 978-3-7315-0016-2

Band 18 Kristina Kreppenhofer
Modular Biomicrofluidics - Mikrofluidikchips im Baukastensystem für Anwendungen aus der Zellbiologie. 2013
ISBN 978-3-7315-0036-0

Band 19 Ansgar Waldbaur
Entwicklung eines maskenlosen Fotolithographiesystems zum Einsatz im Rapid Prototyping in der Mikrofluidik und zur gezielten Oberflächenfunktionalisierung. 2013
ISBN 978-3-7315-0119-0

Band 20 Christof Megnin
Formgedächtnis-Mikroventile für eine fluidische Plattform. 2013
ISBN 978-3-7315-0121-3

Band 21 Srinivasa Reddy Yeduru
Development of Microactuators Based on the Magnetic Shape Memory Effect. 2013
ISBN 978-3-7315-0125-1

Die Bände sind unter www.ksp.kit.edu als PDF frei verfügbar oder als Druckausgabe zu bestellen.

Band 22 Michael Röhrig
Fabrication and Analysis of Bio-Inspired Smart Surfaces. 2014
ISBN 978-3-7315-0163-3

Band 23 Taleieh Rajabi
Entwicklung eines mikrofluidischen Zweikammer-Chipsystems mit integrierter Sensorik für die Anwendung in der Tumorforschung. 2014
ISBN 978-3-7315-0220-3

Band 24 Frieder Märkle
Laserbasierte Verfahren zur Herstellung hochdichter Peptidarrays. 2014
ISBN 978-3-7315-0222-7

Band 25 Tobias Meier
Magnetoresistive and Thermoresistive Scanning Probe Microscopy with Applications in Micro- and Nanotechnology. 2014
ISBN 978-3-7315-0253-1

Band 26 Felix Marschall
Entwicklung eines Röntgenmikroskops für Photonenenergien von 15 keV bis 30 keV. 2014
ISBN 978-3-7315-0263-0

Band 27 Leonardo Pires Carneiro
Development of an Electrochemical Biosensor Platform and a Suitable Low-Impedance Surface Modification Strategy. 2014
ISBN 978-3-7315-0272-2

Band 28 Sebastian Mathias Schillo
Prozessentwicklung für die Automatisierung der Herstellung und Anwendung von hochdichten Peptidmicroarrays. 2014
ISBN 978-3-7315-0274-6

Band 29 Nicole E. Steidle
Micro- and Nanostructured Microfluidic Devices for Localized Protein Immobilization and Other Biomedical Applications. 2014
ISBN 978-3-7315-0297-5

Band 30 Jochen Heneka
Prozessentwicklung eines industrietauglichen Verfahrens zur Fertigung von vereinzelten LIGA-Mikrobauteilen. 2015
ISBN 978-3-7315-0326-2

Band 31 Seoung-Eun Kim
Konzeption und prototypische Fertigung einer nicht-invasiven mikrofluidischen Plattform für die Elektrophysiologie (NIMEP) zur Zellenanlyse. 2015
ISBN 978-3-7315-0378-1

Band 32 Elisabeth Wilhelm
Entwicklung eines mikrofluidischen Brailledisplays. 2015
ISBN 978-3-7315-0385-9

Band 33 Viktor Pinneker
Entwicklung miniaturisierter Aktorsysteme basierend auf magnetischen Formgedächtnislegierungen. 2016
ISBN 978-3-7315-0500-6

Band 34 Ali Caglar Özen
Novel MRI Technologies for Structural and Functional Imaging of Tissues with Ultra-short T_2 Values. 2017
ISBN 978-3-7315-0657-7